SOTHIS PRESS

Sothis Press aims to completely reimagine the public domain book with an emphasis on universal access to captivatingly designed print editions. We believe that technology will enable every single book ever published to be perpetually accessible in beautiful print form. Sothis is beauty in ubiquity.

www.sothis.press
1 Waterhouse Square, London, ECIN 2ST

The Westinghouse E-T Air Brake Instruction Pocket Book

A Complete Work Explaining in Detail the Improved Westinghouse Locomotive Air Brake Equipment, Including Both the No; 5 and the Latest, Perfected No; 6 Style; Contains Examination Questions and Answers, Covering What the E-T Equipment Is; How It Should Be

W. W. Wood

SOTHIS PRESS

ISBN 978-1-0365-2580-4

The Westinghouse E-T Air Brake Instruction Pocket Book is in the public domain.

This edition is published by Sothis Press The company is registered in England and Wales. The Company Number is 14900389.

The Westinghouse E-T Air Brake Instruction Pocket Book

A Complete Work Explaining in Detail The Improved Westinghouse Locomotive Air Brake Equipment, Including both the No. 5 and the Latest, Perfected No. 6 Style

CONTAINS EXAMINATION QUESTIONS AND ANSWERS, COVERING WHAT THE E-T EQUIPMENT IS. HOW IT SHOULD BE OPERATED. WHAT TO DO WHEN DEFECTIVE. NOT A QUESTION CAN BE ASKED OF THE ENGINEMAN UP FOR PROMOTION ON EITHER THE NO. 5 OR THE NO. 6 EQUIPMENT THAT IS NOT ASKED AND ANSWERED.

by W. W. WOOD, Air Brake Instructor

Author of "The Walschaert Locomotive Valve Gear," "Locomotive Breakdowns," etc., etc.

FILLED WITH COLORED PLATES, SHOWING VARIOUS PRESSURES WHICH HELP TO ASSIST THE READER IN UNDERSTANDING THE EFFECT PRODUCED IN THE VARIOUS PHASES OF LOCOMOTIVE AND TRAIN BRAKE OPERATION.

NEW YORK

THE NORMAN W. HENLEY PUBLISHING CO.

132 Nassau Street

1909

PREFACE

Previous to the issuance of this volume there has been a general similarity in all air-brake instruction books, and the E-T Air-Brake Pocket-Book is the first departure to a wholly original field of air-brake instruction.

Since its inception, and until quite recently, there has been practically but one style of air brake for all classes and branches of railroad service, both passenger and freight, and its application to the locomotive has been heretofore the simplest modification of the plain, automatic principle. With the great increase in weight and motive power of the locomotives in general service at the present day, the importance of their braking power has increased enormously, and it is common to hear a locomotive engineer declare that he would rather have one-half of the car brakes of a long freight train out of operation than to have to cut his locomotive brake out of action. The different classes of train service now require different methods of brake operation. The running time of passenger trains has been increased, also, at such a rate that the comparatively modern evolution of the quick-action brake for High-Speed Service has been short-lived, and the demand heeded for a further increase in the stopping power of the brakes of the cars and the locomotives in general passenger service.

Under the circumstances, and to provide for certain other present-day air-brake requirements, it has been found necessary to produce a locomotive brake with an individuality of its own, but perfectly adapted to every requirement of each branch of railroad service; and as the result of much invention, experiment and re-designing, the improved E-T Locomotive-Brake Equipment has been brought forth by the Westinghouse Air-Brake Company, and is now the one standard type of engine- and tender-brake for each and every locomotive, regardless of the service in which it may be placed.

In the E-T equipment the entire apparatus included in the engine- and tender-brake has been reconstructed, and although the *principle* of the common triple valve is used to govern the graduation of the locomotive braking power in like proportion to the calculated power of the car brakes of the train, the general construction of this new equipment is so different from the old that it is practically impossible for a person otherwise skilled in knowledge of the common quick-action and locomotive brakes, to understand the E-T equipment without helpful instruction.

It should be borne in mind, too, that when any person thoroughly understands the E-T brake equipment, he has competent knowledge of the Westinghouse air brake as it is applied to any locomotive in this country, for, while a knowledge of the common automatic air brake is helpful in the study of the E-T equipment, a thorough understanding of the latter embraces all that has gone before, in addition to the improvements that make this the accepted air-brake standard for all locomotives of the future. It is therefore of the greatest importance that locomotive engineers in particular, but all those as well who have anything to do with the maintenance of locomotive equipment, or the supervision of enginemen or air-brake repair men, shall become well informed concerning the improved E-T locomotive-brake equipment.

Constant attendance at an air-brake instruction car, until the new type of brake can be perfectly understood, is a privilege not within the reach of the many who desire the knowledge; and for those who may receive such direct instruction, good literature on the subject is as helpful as the text-book is to the college student. With the object in view of aiding all those who desire to become proficiently acquainted with the E-T locomotive-brake equipment, this book has been written. Its production was delayed until the equipment had reached practical perfection in the No. 6, its latest, improved style, upon which the text and illustrations are mainly based, although the preceding, No. 5, style of the equipment is fully described, in so far as it differs from the perfected type.

While the main portion of this book is designed for the assistance of those who will not be satisfied with anything short of a complete and thorough understanding of the whole E-T locomotive-brake equipment, a series of Questions and Answers has been appended for the benefit of those who wish to become posted in the material details, only, of this equipment, in order to qualify for an examination thereon; and this section is so complete in itself, that it is recommended to Travelling Engineers and Air-Brake Inspectors and Instructors as a standard form of Examination Questions on the No. 5 and No. 6 E-T equipments.

The illustrations are wholly original, the scheme of giving each zone of air pressure its distinctive color causing the plates to appear self-explanatory to a great extent. The text is also original, except that the descriptive language used by the air-brake company in its literature is occasionally employed in reference to certain parts of the apparatus, as examples of brevity that can not be improved upon.

This book is, as the name implies, a text-book and reference work on the E-T LOCOMOTIVE-BRAKE EQUIPMENT; and the accessories that are the same as were used in connection with the common automatic engine- and tender-brake, such as the air pump, the train air-signalling equipment, and the foundation-brake gear, are not touched upon, although in a later edition a full description of those parts is contemplated.

THE AUTHOR.

June, 1909.

TABLE OF CONTENTS

Contents

THE WESTINGHOUSE

No. 6

E-T Locomotive Brake Equipment

The improved, Westinghouse, locomotive air-brake equipment (the term *locomotive brake* meaning the combined braking apparatus of the engine and tender, which in this system works as a unit), is denominated by the symbols E-T (engine-tender), and the perfected equipment is designated as the No. 6. The No. 5 style of this brake was brought out in 1905, and has been quite generally applied to locomotives built since that date until succeeded by the more perfect design. While the No. 6 does not differ greatly from the preceding style, the points of difference are important enough to warrant the adoption of the latest construction for all locomotives that will be built in the future, and this, the No. 6 E-T equipment, has been selected as the subject of this book of instruction; after it has been fully described, and illustrated, all necessary reference will then be made to the differences embodied in the No. 5, or older, style.

Foreword

The essential idea in the production of the E-T equipment is to furnish a dependable automatic, locomotive brake—which the simple automatic type was not, as it was hard to keep the brake cylinders even reasonably free from pressure leakage. The secondary, *straight-air*, or "independent" brake on the locomotive had become a necessity, and, together with other improvements and attachments demanded by the service in modern train braking, the older system became complicated and erratic. A radical change has been made, and a new type of automatic brake for the locomotive evolved: the E-T, which consists of considerably less apparatus than the former "combined automatic and straight-air brake," while possessing all the advantages of the latter and several other important ones which are necessary in connection with modern locomotive brake appliances.

There being but one equipment (and not requiring different sizes of valves to conform to the several sizes of brake cylinders), it may be applied to any locomotive whether used in high-speed passenger, double-pressure control, ordinary passenger or freight, or any kind of switching service, without change or special adjustment of the brake apparatus. All valves are so designed that they may be removed for repairs and replacement without disturbing the pipe joints.

The locomotive brakes may be used with or independently of the train brakes, and this without regard to the position of the locomotive in the train. They may be applied with any desired pressure between the minimum and the maximum, and this pressure will be automatically maintained in the locomotive-brake cylinders regardless of leakage from them and of variations in piston travel, undesirable though these defects are, until released by the brake valve. They can be graduated on or off with either the automatic or the independent brake-valves; hence, in all kinds of service the train can be handled without shock or danger of parting, and in passenger service smooth, accurate stops can be made with greater ease than was heretofore possible.

PARTS OF THE EQUIPMENT.

(See Fig. 1.)

1. The AIR PUMP to compress the air.

2. The MAIN RESERVOIR, in which to store and cool the air and collect water and dirt.

3. A DUPLEX PUMP-GOVERNOR to control the pump when the pressures are attained for which it is regulated.

4. A DISTRIBUTING VALVE, and small double-chamber reservoir to which it is attached, placed on the locomotive to perform the functions that have heretofore devolved on the triple valves, auxiliary reservoirs, double-check-valves, high-speed reducing valves, etc.

5. Two BRAKE-VALVES, the AUTOMATIC to operate locomotive and train brakes, and the INDEPENDENT to operate locomotive brakes only.

6. A FEED VALVE to regulate the brake-pipe pressure.

7. A REDUCING VALVE to reduce the pressure for the independent brake-valve and for the air-signal system when used.

8. Two DUPLEX AIR-GAUGES; one, to indicate equalizing-reservoir and main-reservoir pressures; the other, to indicate brake-pipe and locomotive-brake-cylinder pressures.

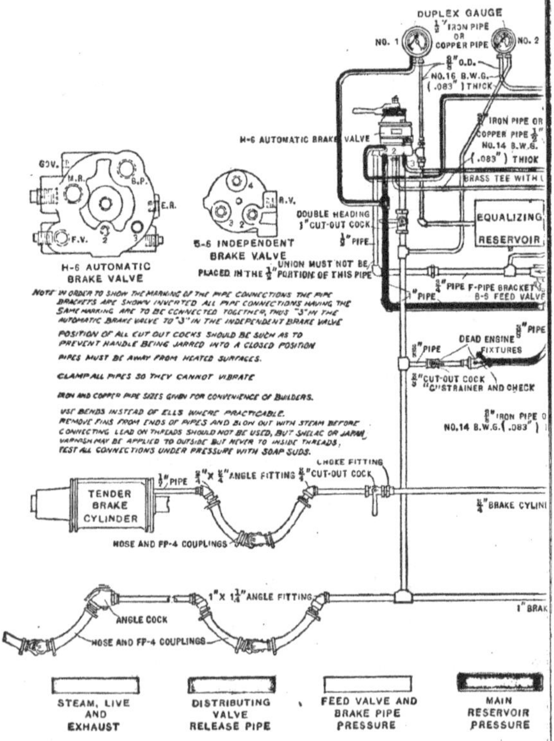

Fig. 1.—Piping Diagram, No. 6 E T Locomotive-Br
routi

Fig. 1

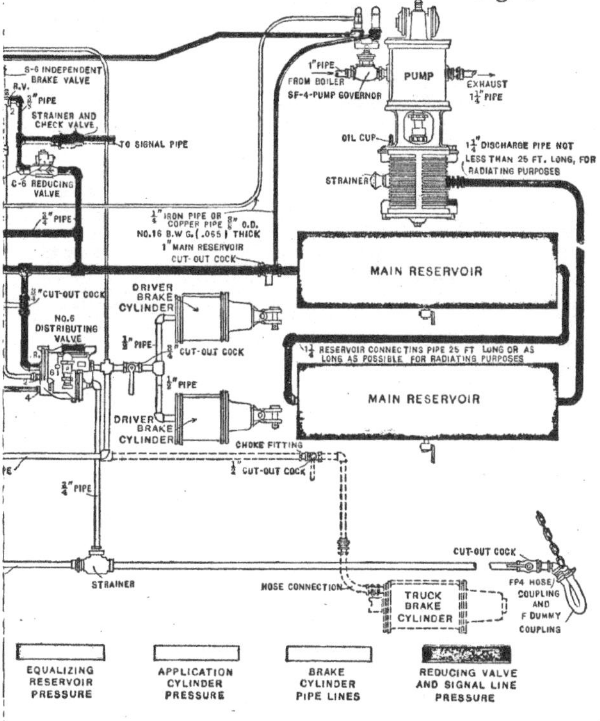

Equipment. Colors showing open communications, and of pressure.

Fig. 1

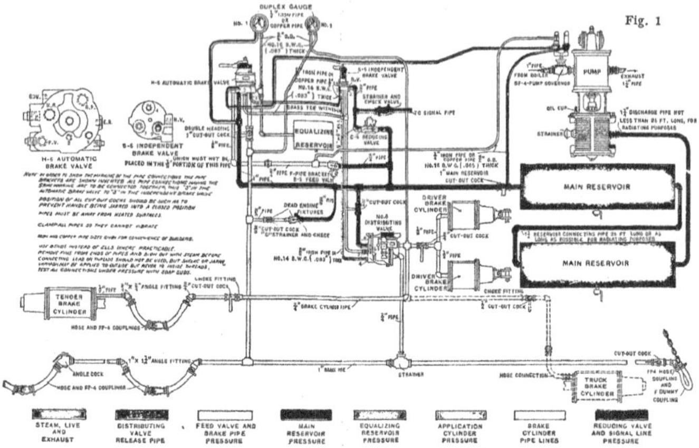

Fig. 1.—Piping Diagram, No. 6 E T Locomotive-Brake Equipment. Colors showing open communications, and routings of pressure.

9. DRIVER, TENDER, and TRUCK-BRAKE CYLINDERS, CUT-OUT COCKS, AIR STRAINERS, HOSE COUPLINGS, FITTINGS, etc., incidental to the piping. (The 1-inch Cut-out Cock in Brake Pipe directly beneath the automatic brake-valve (Fig. 1) is designated the DOUBLE-HEADING COCK).

NAMES OF PIPING.

(See Fig. 1. Notations thereon as to pipe connections.)

In the color scheme of Fig. 1, it is not implied that the indicated pressures are in all of the several pipes at the same time; the intention is to explain the routing of the air flow, and the zone that may be occupied by each pressure.

In subsequent charts, the coloring will indicate the pressures that are contained in the pipes, chambers, etc., during the particular phase of action represented by the individual plate.

DISCHARGE PIPE: Connects the AIR PUMP to the first MAIN RESERVOIR.

CONNECTING PIPE: Connects the two MAIN RESERVOIRS.

MAIN-RESERVOIR PIPE: Connects the second MAIN RESERVOIR to the AUTOMATIC BRAKE-VALVE, DISTRIBUTING VALVE, FEED VALVE, REDUCING VALVE, and PUMP GOVERNOR.

FEED-VALVE PIPE: Connects the FEED VALVE to the AUTOMATIC BRAKE-VALVE.

EXCESS-PRESSURE GOVERNOR PIPE: Connects the FEED-VALVE PIPE to the EXCESS-PRESSURE HEAD of the PUMP GOVERNOR.

REDUCING-VALVE PIPE: Connects the REDUCING VALVE to the INDEPENDENT BRAKE-VALVE, and to the SIGNAL SYSTEM.

BRAKE PIPE (Formerly "*train line.*"): Connects the

AUTOMATIC BRAKE-VALVE with the DISTRIBUTING VALVE and all TRIPLE VALVES on the cars in the train.

BRAKE-CYLINDER PIPE: Connects the DISTRIBUTING VALVE with the DRIVER, TENDER and TRUCK-BRAKE CYLINDERS.

APPLICATION-CYLINDER PIPE: Connects the APPLICATION CYLINDER of the DISTRIBUTING VALVE with the INDEPENDENT and AUTOMATIC BRAKE-VALVES.

DISTRIBUTING-VALVE RELEASE PIPE: Connects the APPLICATION-CYLINDER exhaust port of the DISTRIBUTING VALVE to the AUTOMATIC BRAKE-VALVE through the INDEPENDENT BRAKE-VALVE.

ARRANGEMENT OF APPARATUS, PIPE CONNECTIONS, AND GENERAL ROUTE OF PRESSURE.

Referring to Fig. 1, the air compressed by the pump passes as usual to the main reservoirs and the main-reservoir pipe; this pressure is indicated by the red color, and in this diagram all parts of one color are in open communication with each other. The main-reservoir cut-out cock is of the 3-way style, and when closed will cut off the air from the main reservoir, and vent the pressure from the main-reservoir pipe and all other pipes, valves, etc., in connection thereto—the main brake pipe as well, unless the "1-inch cut-out cock" beneath the automatic brake-valve should first be closed; besides this, the brake-valve handle should be placed in **release position** to prevent the slide valve of the feed valve, and the rotary valve of the brake-valve being lifted from their seats; any part of the apparatus, except the governor, may then be removed without the necessity of stopping the pump and emptying the main reservoir. The end of the 3-way cock toward the main reservoir is tapped for a connection to the *high-pressure head* of the pump governor, and will restrain the pump from working up any higher pressure than the desired maximum in the main reservoir while repairs to the apparatus are being made.

Beyond the main-reservoir (3-way) cut-out cock, the main-reservoir pipe has four branches; one of which runs to the automatic brake-valve to supply the pressure when the brake-valve is in **release position** for the quick recharge of the brake pipe necessary in releasing the train brakes; one branch runs to the feed valve, which reduces the pressure that the automatic brake-valve will take for the brake-pipe supply to 70 pounds—in all branches of service except with the high-speed passenger brake when it is adjusted to 110 pounds—and the feed-valve pipe delivers this reduced pressure to the automatic brake-valve, through which it is supplied to the brake pipe in the **running** and **holding positions;** one branch of the main-reservoir pipe leads to the reducing valve—adjusted generally at 45 pounds—which regulates the pressure used in the air-signal system, and by the independent brake-valve; and one branch to the distributing valve through which the main-reservoir air is automatically supplied in graded amount to the brake cylinders of the engine and tender at all applications, by the independent or automatic brake-valves, at service and emergency reductions. As a result, the automatic brake-valve receives air from the main reservoir in two ways, one direct and the other through the feed valve.

The feed-valve pipe from the feed valve to the automatic brake-valve has a branch to the top of the excess-pressure head of the duplex pump-governor, and the reasons for this connection will appear in the explanation of the SF-4 pump-governor.

The reducing-valve pipe leading from the reducing valve to the independent brake-valve, has a branch-pipe connection with the train air-signal system, when used. In this branch pipe is placed a combined strainer, check-valve, and choke-fitting: the check-valve to prevent return flow of the signal-line pressure when an application by the independent brake-valve is made, and which would cause the air-signal whistle to sound; the strainer to prevent dirt from lodging in the check-valve; and the choke-fitting to so reduce the rapidity of air supply to the signal line that the opening of the car discharge-valve can reduce the pressure in the line and cause the air whistle to blow.

The distributing valve has five pipe connections (see Figs. 2 A and 2 B), made through the end of the double-chamber reservoir, three on the left and two on the right; only one of them (the middle one on the left) enters either of the chambers of the reservoir, and they are all directly related to the valve section, being ported through the large, round gasket between the faces of the reservoir and valve sections of the distributing valve. The pipe connections being made to the reservoir section, it enables the removal of the valve section for exchange or repair without having to disconnect any pipe

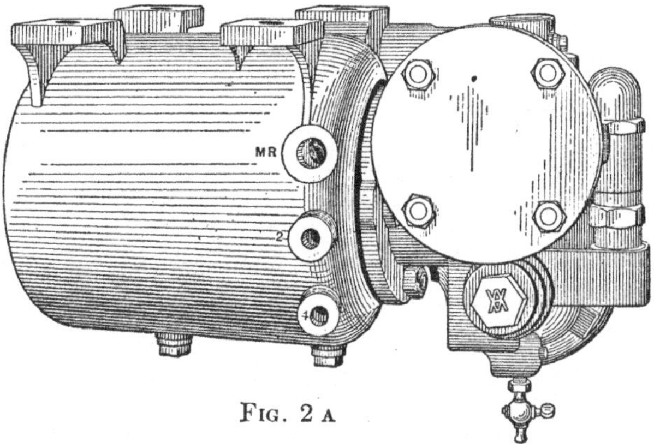

FIG. 2 A

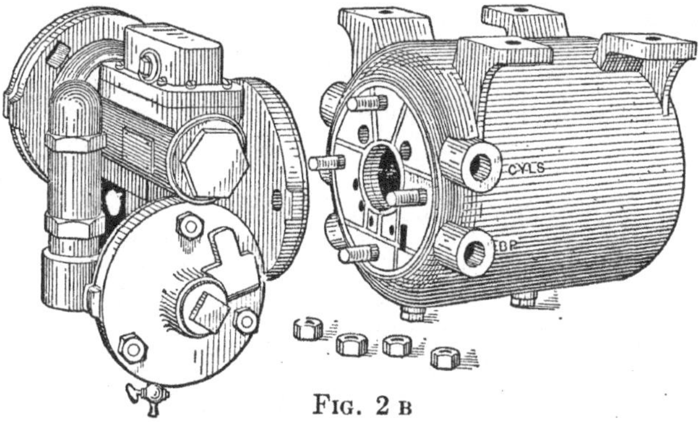

FIG. 2 B

FIG. 2 A.—No. 6 Distributing Valve, Left Side, showing the Pipe Connections. MR, to main reservoir; 2, application-cylinder pipe; 4, distributing-valve release-pipe.

FIG. 2 B.—No. 6 Distributing Valve, Right Side. CYLS, brake-cylinder pipe (branching to all brake cylinders of engine and tender); BP, distributing-valve branch of the brake pipe.

unions. Of the three pipes on the left, the upper is the supply from the main reservoir; the intermediate is the application-cylinder pipe, leading to the independent and the automatic brake-valves; and the lower is the distributing-valve release pipe, leading through the independent brake-valve, when the handle is in **running position,** to the automatic brake-valve and through it, also, when in **running position,** to the atmosphere, as shown in Fig. 1 by the continuity of color. Of the two on the right, the lower is the brake-pipe branch connection, and the upper is the brake-cylinder pipe, branching to *all brake cylinders on the engine and tender.* In this pipe are placed cocks for cutting out, individually, the brake cylinders when necessary, one each for the tender and engine truck-brake cylinders, and one cock to cut out both driver-brake cylinders at once, as it is inadvisable to operate the driver brake on one side of the engine only; and in the engine truck and tender brake-cylinder cut-out cocks are placed choke-fittings to prevent serious loss of main-reservoir air and the release of the other locomotive brakes during a stop, in case of burst brake-cylinder hose. Each one of the pieces of the E-T equipment referred to in this description of the piping arrangement will be explained in detail, and with accompanying illustrations, further along.

The TWO DUPLEX AIR-GAUGES (see Figs. 3 A and

3 B) are connected as follows: Gauge No. 1; red hand to main-reservoir pipe under the automatic brake-valve; black hand to gauge-pipe tee of the automatic brake-valve; this gauge is piped correspondingly the same as the original duplex gauge of the automatic-brake equipment with which we have been familiar

FIG. 3 A FIG. 3 B

FIG. 3 A.—Large Duplex Gauge. (Corresponding to the one duplex gauge used in former equipments.)

FIG. 3 B.—Small Duplex Gauge. (Corresponding to the *brake cylinder gauge* used in former equipments of the E-T brake, but with a second pointer added to show brake-pipe pressure direct.)

for years, but the inscription on the dial is more strictly accurate, for instead of the legend "black hand *train line*"—meaning *brake-pipe* pressure—it now says BLACK HAND EQUALIZING RESERVOIR, which pressure it records and always has recorded. Gauge No. 2; red hand to the brake-cylinder pipe; black hand to the brake pipe below the double-heading cock; thus, for the first time, in connection with the *equalizing-discharge* automatic brake-valve a gauge is supplied that *directly* indicates the pressure in the brake pipe at all times, regardless of the position of the brake-valve handle, and is of special benefit on the secondary engines in double-heading whose double-heading cocks are closed.

The amount of reduction to be made during an automatic application, however, is indicated by the black hand of gauge No. 1, as the E-T equipment does not require any change in the governing of automatic-brake applications at either service or emergency reductions.

The automatic brake-valve connections, other than those already mentioned, are the brake pipe, the pipe to the equalizing reservoir (sometimes referred to as the *chamber D reservoir*), and the lower connection to the excess-pressure head of the pump-governor.

Before beginning on a detailed description of the operation, and the construction of the operating parts, of the No. 6 E-T equipment, it is desired that it shall be positively understood that the underlying principles governing its action are just the same as those of all previous equipments of the automatic air-brake. Instead of a triple valve and auxiliary reservoir for each of the engine and tender equipments the parts mentioned are dispensed with, and the DISTRIBUTING VALVE is made to take the place of the triple valve, but one being used to control the supply of pressure to, and its discharge from, all brake cylinders of the engine and tender; while the presence of the main reservoir on the engine has been taken advantage of for direct brake-cylinder air supply, to supplant the auxiliary reservoir.

AN EXPLANATION OF THE PRINCIPLE THAT GOVERNS THE GRADUATED OR COMPLETE APPLICATION, AND THE RELEASE, OF ALL TYPES OF THE AUTOMATIC AIR BRAKE.

The first "continuous train brake" to come into general use was of the *straight-air* type, invented by George H. Westinghouse, and, succeeding the hand brake, was comparatively successful on a limited number of cars, and *as long as everything worked all right.* The straight-air brake can be, and is, absolutely reliable, but only when used on the same vehicle from which the actuating pressure is originally supplied and that carries the main operating brake-valve. The automatic brake as it exists to-day is due to the production by Mr. Westinghouse of the *triple valve*, which is the one essential part of each air-brake unit (car or locomotive equipment), and no automatically acting brake can be conceived to work with compressed air that does not make use of the principle of the triple valve.

THE TRIPLE VALVE.—Nominally, the Westinghouse E-T locomotive-brake equipment does not include a triple valve; really, however, that portion of the distributing valve called the *equalizing valve* performs the functions of a triple valve, and is necessary

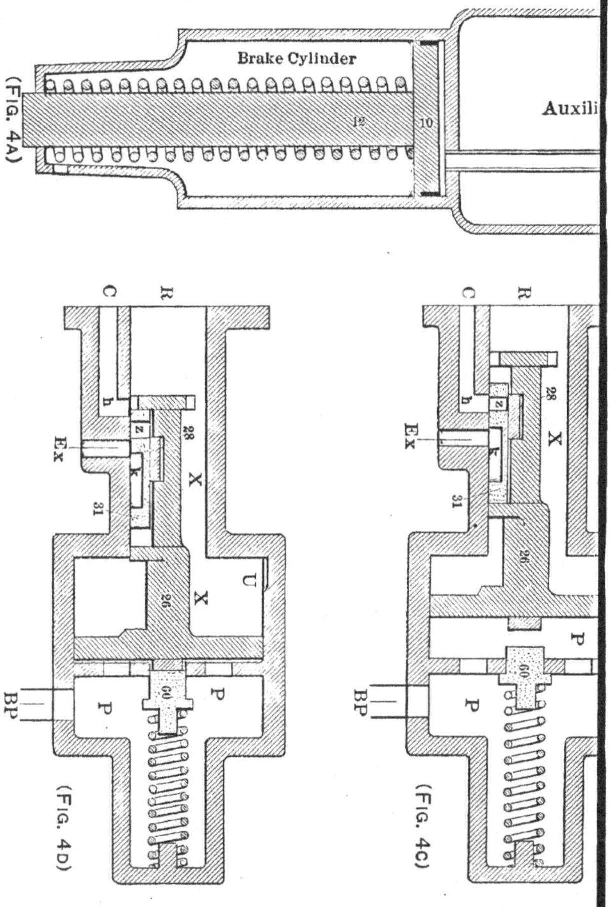

FIG. 4 A.—Triple Valve in Release Position, with Auxiliary R
FIG. 4 B.—Triple Valve in Service-Application Position.
FIG. 4 C.—Triple Valve in Service-Lap Position.
FIG. 4 D.—Triple Valve in Emergency-Application Position.

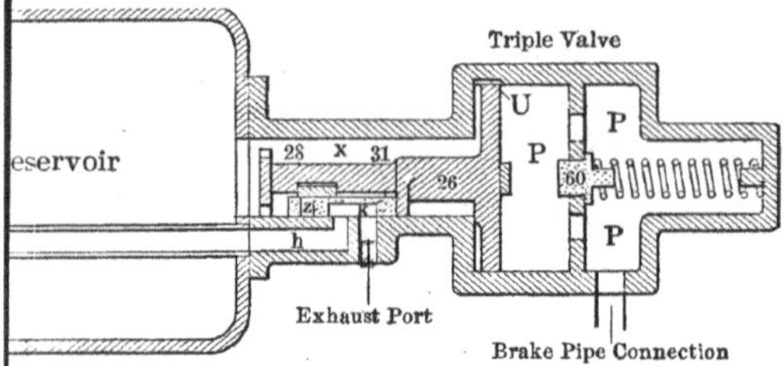

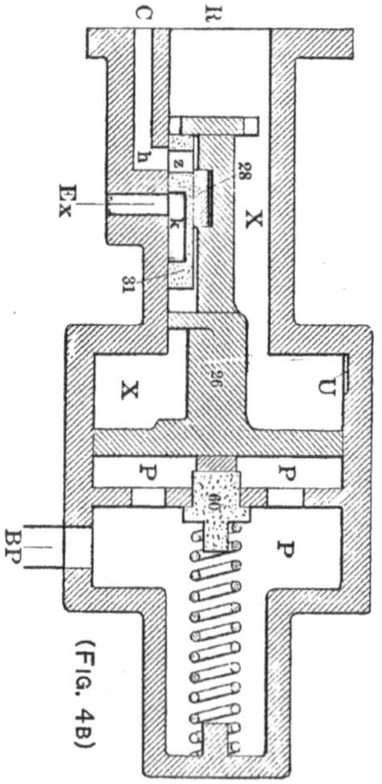

(FIG. 4B)

COLOR KEY FOR FIGS. 4A, 4B, 4C, 4D.

BRAKE PIPE PRESSURE.

AUXILIARY RESERVOIR PRESSURE.

BRAKE CYLINDER PRESSURE.

ATMOSPHERIC AIR. PRESSURE.

oir and Brake Cylinder. Charged. Diagrammatic figure.

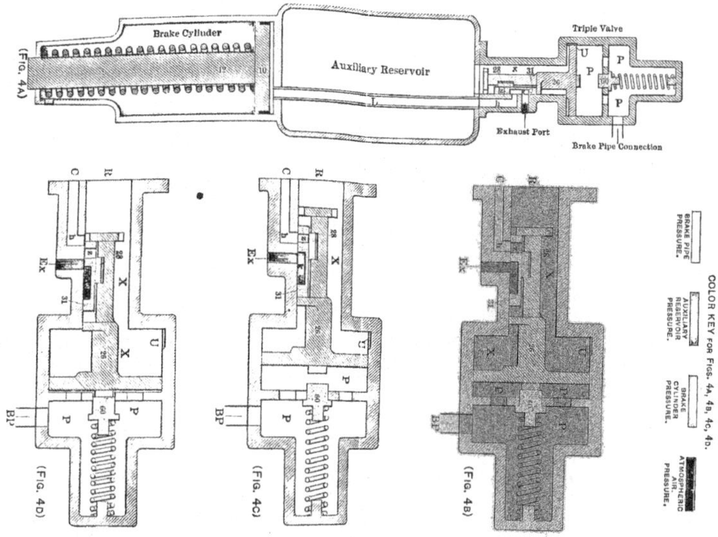

FIG. 4 A.—Triple Valve in Release Position, with Auxiliary Reservoir and Brake Cylinder. Charged. Diagrammatic figure.
FIG. 4 B.—Triple Valve in Service-Application Position.
FIG. 4 C.—Triple Valve in Service-Lap Position.
FIG. 4 D.—Triple Valve in Emergency-Application Position.

to correctly graduate the locomotive-braking power at service applications, and to so operate that the power of the locomotive brake will be proportionately the same as that of the properly working car brakes during the different grades of application. In fact, it does all that a triple valve has done, and performs other additional functions that are individual to the E-T type of brake equipment.

It is necessary, therefore, to be familiar with the operation of a plain triple valve before it is possible to satisfactorily understand the distributing valve; and as the latter is as essentially the vital part of the E-T equipment as the triple valve is of the plain automatic brake, the working of a triple valve of the simplest design of construction will be exemplified, and will be followed by an ideal, or diagrammatic, illustration and explanation of the principal features of the distributing valve for comparison, and which will make it clear that the original theory of the automatic brake has not been departed from, but only strengthened.

Fig. 4 A represents a triple valve with the parts in the *release position*, together with the auxiliary reservoir, and brake cylinder and piston; the triple valve embodies the controlling mechanism, the auxiliary reservoir is to contain the pressure ready to be used in the application of the brake, and the brake cylinder is the place where the power of the compressed air is made to transmit its force to the brake shoes through the *foundation brake*, or rigging, by acting against the movable piston contained within the cylinder. Figs. 4 B, 4 C, and 4 D show the same triple valve with the operating parts in the *service-application*, *service-lap*, and *emergency-application* positions, respectively, but without the auxiliary reservoir and brake cylinder—the points of connection therewith being indicated—as their duties will be sufficiently understood after the explanation of the first diagram, Fig. 4 A.

Alluding to Fig. 4 A, the triple-valve piston, 26, contains a packing ring (not shown), to make a practically air-tight joint between the spaces on each side of it, and, seated between the shoulders of the stem of the piston is a slide valve, 31, that follows the piston, but with a certain amount of slack, or lost motion, which the piston must take up in either direction before the slide valve moves; the top of this slide valve forms the seat for another one of smaller design but similar in some respects, called the graduating valve, No. 28, which is combined with the piston and moves regularly with it; while piston 26 is air-tight on its edges, in the release position it is necessary that there shall be a slender opening between chambers *p* and *x*, and to serve this purpose a small groove, *u*, is cut in the wall of the triple-valve cylinder, but made so short that the triple piston must be fully in the release position, as here shown, in order that air will feed through it. The connections of the triple valve are as follows: The brake-pipe connection, as indicated on the plate, is to a branch of the main pipe that extends throughout the train from the engineer's brake-valve, and which is constantly charged with pressure from the main reservoir, that the feed valve maintains at 70 pounds while the engineer's valve is in *running position.* The brake pipe was formerly known as the "train line," and the duty of its contained pressure is two-fold: to maintain the air charge of the auxiliary reservoir on each car, and to be the medium by which the engineer can operate the triple valves on all of the cars simultaneously—reducing the pressure to apply the brakes, and increasing it to release them; the connection of the triple valve with the auxiliary reservoir is plainly seen in the opening through the front head of the reservoir to which the triple valve is attached; the tube L, extending through the auxiliary reservoir, furnishes the connection between the triple valve and the brake cylinder; the exhaust port to the atmosphere is indicated on the plate, and with the triple valve in release position, as shown, the pressure space of the brake cylinder is open to the atmosphere through the tube L, port *h* that opens into the seat of the triple-slide-valve, cavity *k* in the face of the slide valve, and the exhaust port. When pressure is released from the brake cylinder, the large spring around piston rod, 12, holds the brake piston, 10, against the pressure head of the cylinder; the extension of the left, or non-pressure, head of the brake cylinder is to permit the nesting of the release spring for protection in case the piston should ever be permitted to travel the full, possible distance of the cylinder; but the slack in the brake rigging should be taken up enough that at a full-on application the piston will not travel much over one-half of its full stroke.

Air pressure from the brake pipe enters the triple valve as shown, filling chambers *p*, *p*, and passing through the feed groove, *u*, charges chamber *x* and the auxiliary reservoir, but slowly, under a full head of brake-pipe pressure at the rate of about one pound of reservoir charge per second; when the auxiliary reservoir has become charged to 70 pounds, or equal to the brake-pipe pressure, no movement of the triple-valve parts will automatically occur, as the piston is in a perfect equilibrium of brake-pipe pressure on one side and auxiliary-reservoir pressure on the other.

Fig. 4 B—SERVICE-APPLICATION POSITION: A reduction of brake-pipe pressure having been made—say of 10 pounds—the pressure in chamber *p* is now 60 pounds, and as the feed-groove, *u*, is too small to permit the auxiliary air to equally reduce by back flow through it, the 70-pounds pressure in chamber *x* forces the triple piston, 26, to the right until its knob comes into contact with the spring head, 60, at which point the piston's movement ceases, with the triple valve in the *service-application position,* as shown; the first movement occurs when the brake-pipe pressure has been reduced by 2 to 4 pounds, when the piston responds by moving far enough for the shoulder on the end of the stem to strike against the left end of slide valve 31, and if the brake-pipe pressure reduces slowly—as is always the case when a service application is made from a long train line—there will be a slight pause of the triple-valve piston at this point, due to the frictional resistance of the slide valve from the air pressure upon it, but the feed-groove, *u,* will have been closed, thereby trapping the auxiliary-reservoir air against any back flow, and graduating valve 28 will have been drawn from over port *z* in the slide valve; when the brake-pipe reduction has amounted to from 5 to 7 pounds, the resistance of slide valve 31 will have been overcome and the triple piston will complete its service travel, carrying the slide valve with it and bringing port *z* into register with port *h* in the slide-valve seat; the auxiliary-reservoir air in chamber *x* now flows through port *z* to port *h,* and enters the tube L, through the auxiliary reservoir, that carries it to the brake cylinder, where, acting upon the brake piston, the air pressure accomplishes the application of the brake.

The *automatic graduation* of the pressure supplied to the brake cylinder is the paramount feature of the triple valve. It may be asked at this point—why, if graduating port *z* is fully opened to port *h*, does not the brake apply full-on, by the auxiliary-reservoir air continuing to feed to the brake cylinder until their pressures are equal? And the answer to this question explains the "secret" of the triple valve—gives the reason for its automatic graduation of braking power: As the result of a 10-pound reduction the brake-pipe pressure stands at 60 pounds; the auxiliary-reservoir air ("*you can't have your penny and spend it, too*"), in supplying the brake cylinder, drops in pressure until it becomes just a trifle *less than the brake-pipe pressure*, and when that occurs the superior pressure acting upon the triple-valve piston is on the right—the brake-pipe side—of it, and although the difference in pressures may only be a matter of ounces there is but little frictional resistance to piston 26's leftward movement until it slides the short distance necessary to bring the other shoulder of the piston's stem against slide valve 31, and then the frictional resistance of the slide valve is encountered and the triple piston is again stopped; the slight, leftward movement of piston 26 closed the graduating valve, 28, cutting off further supply to the brake cylinder and arresting the fall of auxiliary-reservoir pressure when it has become just enough weaker than that of the brake pipe to permit of the back lash of the piston, but not weak enough to permit movement of the slide valve; the triple valve has then assumed the position of *service lap*, as shown in Fig. 4 C.

It should be stated that the graduating spring that offers an effective resistance to the movement of the triple-valve piston beyond the *service-application position* is not absolutely necessary in order that the piston shall stop at that point; a triple valve may work very well without the graduating spring; it may be removed, and if the brake-pipe pressure is not reduced more rapidly than the rate provided for in the equalizing-discharge port of the engineer's brake-valve, and if the triple valve is not sticking and sluggish in movement as the result of dirt and lack of lubrication, the service movement will be accomplished just as well and the triple-valve piston will stop in exactly the same position as though the resistance of the graduating spring had been interposed; the reason is that when the triple valve has reached the service-stop position the auxiliary-reservoir pressure that moved it thus far begins to reduce through the graduating port to the brake cylinder, and this fall of the motive force will be as rapid as, or more so than, the reduction of the brake-pipe pressure, with the result that the air pressures on both sides of the triple-valve piston are nearly equalized, and the frictional resistance of the slide valve overcomes any slight balance of application force, and the whole mechanism is halted until the back lash to *service lap* occurs.

Fig. 4 D—EMERGENCY-APPLICATION POSITION: Port *h* in the slide-valve seat is of greater area than appears, as, instead of being a circular hole, it extends transversely across the valve seat to nearly the width of the slide valve, and only its narrowest diameter appears; but at service application the pressure can not flow through it any faster than the smaller, round, graduating port, *z*, can supply, which is a rate of flow desirable for service action of the brake; when it is desired to stop quickly, however, the full capacity of the large port in the seat is demanded to supply the reservoir air to the brake cylinder rapidly enough—say in "spotting" the engine on the turntable, at a water column, or in case of emergency on the road—and when such an occasion arises, the engineer's brake-valve being thrown to the *emergency-application position*, the sudden and heavy reduction of brake-pipe pressure induces the movement of the triple-valve piston to the right in the same manner as explained in connection with Fig. 4 B, except that the stroke is quicker, and instead of the piston being halted by the graduating spring the latter is compressed and the piston completes its full travel, assuming the *emergency application position* as represented by Fig. 4 D, in which it is seen that the slide valve, 31, is drawn completely off of port *h* in the seat, exposing the full opening of the latter for the more rapid passage of the auxiliary-reservoir pressure to the brake cylinder.

Referring again to *service application:* In order to apply the brake with full-service force it is only necessary to reduce the brake-pipe pressure to the same figure at which the auxiliary reservoir and brake cylinder will equalize; with an initial pressure of 70 pounds in the former they will not always equalize at the same figure, and this is due to the variation in volumes of the pressure spaces in the brake cylinders of a train, this pressure space being greater or less as the brake piston may have a longer or shorter travel; the length of piston travel is proportionate to the amount of slack in the brake rigging—the longer the piston travel the greater the space that must be filled in the cylinder; and with increased expansion of the air there is decreased pressure; the piston should never travel much more than one-half of its full stroke, and if properly adjusted, the auxiliary reservoir and brake cylinder should equalize at 50 pounds per square inch, and to secure this equalization calls for a 20-pound brake-pipe reduction (from 70 pounds to 50 pounds); in regular train service, the equalization will give brake-cylinder pressures anywhere between 45 and 55 pounds, due to minimum and maximum piston travels, and the intermediate; and as the piston travel on all cars in the train is not known to the engineer, when a full-on service application is to be made it is necessary to reduce the brake-pipe pressure 25 pounds.

OPERATION OF BRAKE RELEASE: As the *reduction* of brake-pipe pressure influenced the triple valves to cause an application of the brakes, it is through the medium of the brake-pipe air that the engineer releases them, which he does by *increasing* the pressure; but, whereas an automatic application of all brakes may be made at any point in the train by reducing the brake-pipe pressure, the engineman, only, can *automatically* release them, as the main reservoir carrying the high releasing-pressure of great volume is located on the engine; if, however, it is required to release brakes on cars that are detached from the engine, or in case of a very long train when the main reservoir may not be pumped-up to a pressure sufficiently high to release all brakes, those remaining applied are said to be "stuck," and can be released by the trainmen "bleeding" each one individually.

To automatically release the brakes, the engineer's brake-valve is placed in the *release position* which permits the high pressure of the main reservoir to flow to the brake pipe, and the increased pressure entering the triple-valve chambers, *p*, *p*, becomes higher than that of the auxiliary reservoir, and piston 26 will be moved its full traverse to the left, to the Release and Charging Position as shown in Fig. 4 A, in which feed-groove *u* is open, through which brake-pipe air begins recharging the auxiliary reservoir; cavity *k* in the face of slide-valve 31 now connects port *h* with the exhaust port in the slide-valve seat, thus opening a way for the brake-cylinder pressure, flowing back through the tube L, to escape to the atmosphere, after which, the large releasing spring around piston rod 12 pushes brake piston 10 to its release position, as shown, against the pressure head of the cylinder.

To bleed the brakes, the trainman opens the *release valve* (not shown) on the auxiliary reservoir, thus reducing its pressure until it is less than that of the brake pipe when the triple valve will be moved to *release position* in the same manner as explained in reference to *automatic release*, and by the same reason —*i.e.*, the brake-pipe pressure is become greater than that of the auxiliary reservoir; but, to effect the release in this way, the auxiliary reservoir of every applied brake in the train must be individually "bled"; if no pressure remains in the brake pipe, the triple valve will be moved to the position shown in Fig. 4 B by the force of the graduating spring when the auxiliary pressure has been largely reduced by bleeding, after which the brake-cylinder pressure will flow back into the auxiliary reservoir through tube L and ports *h* and *z*, and to the atmosphere through the release valve; in the latter case, the release valve must be kept open until all of the pressure in the auxiliary reservoir and brake cylinder has been discharged.

RUDIMENTS OF THE DISTRIBUTING VALVE.

Owing to its automatic character, the unimproved air-brake can not be depended upon to stay applied for any considerable length of time, as the auxiliary reservoir—the local base of pressure supply—is cut off from recharge by the act of applying the brake, and the inevitable piston packing-leather leakage may be expected to waste away the air from the brake cylinder and auxiliary reservoir; but the locomotive—the heaviest unit of the train, and equipped with a brake equaling in calculated power the brakes of several modern freight cars—carries the main-reservoir pressure, and in the E-T type of locomotive-brake equipment a way has been found to supply pressure to the engine and tender-brake cylinders directly from the main reservoir and yet to retain the automatic action and brake-pressure-graduation in harmony with the triple-valve-operated cars of the train.

The **distributing valve** is the central figure of this new equipment, and before taking up the description and explanation of its mechanism in technical detail, a diagrammatic figure will be used to exemplify the principle on which the brake-cylinder pressure is supplied and the automatic graduation of same is performed.

In Fig. 5, we have a triple valve precisely similar to

the one described, but here denominated the *equalizing valve*, an auxiliary reservoir changed in name to *pressure chamber*, and a sealed vessel containing the same volume as would be in the pressure end of the brake cylinder of Fig. 4 A with the brake piston moved out to its normal travel, but called the *application chamber;* and these, with the addition of the case containing the small piston and valve at the left of the application chamber, constitute the application features of the distributing valve in an ideal form.

The same explanation as given of Figs. 4 A, 4 B, 4 C, and 4 D will apply to the action of the equalizing valve as the result of brake-pipe pressure reductions and recharge; full main-reservoir pressure enters and is contained in chamber *a*, as indicated, and serves to hold application valve 5 seated; from chamber *b*, a pipe leads and branches to all brake cylinders of the engine and tender; therefore, application piston 10 has whatever pressure may be in the locomotive-brake cylinders on one side of it, and the pressure of the *dummy-brake cylinder*, or application chamber, on the other side.

If a brake-pipe reduction of 10 pounds should be made, the equalizing valve operating as a triple valve will permit air to flow from the pressure chamber to the application chamber until the pressure of the former has been also reduced to a fraction less than that

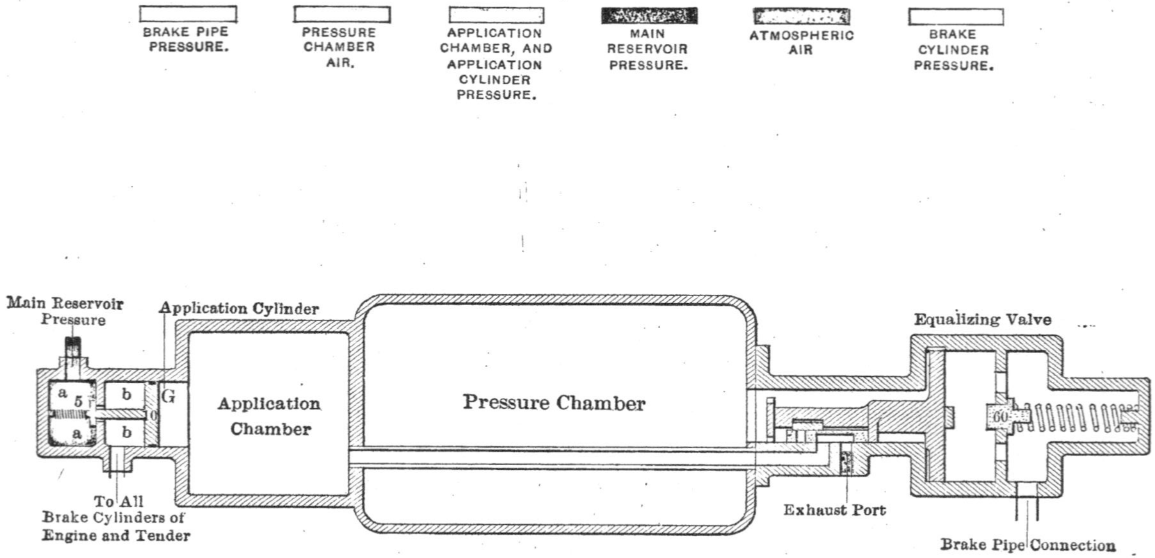

FIG. 5.—Diagrammatic Figure of Distributing Valve. Equalizing feature, at right, showing its similarity in action to a triple valve; and an ideal design of the application feature, at left, to illustrate the method of graduating the pressure supply to brake cylinders from the main reservoir.

in the reduced brake pipe; it will be recalled that in response to a 10-pound brake-pipe reduction the pressure built up in the brake cylinder should be about 25 pounds, *if the piston travel is correct;* but in this arrangement there is no variable-traveling piston that can alter the result of the pressure expansion, and following a 10-pound reduction the application chamber *will have 25 pounds pressure;* this 25 pounds acting upon application piston 10, in application cylinder G, will force it to the left, unseating valve 5 and permitting main-reservoir air from chamber *a* to enter chamber *b* and pass from there to the brake cylinders of the engine and tender; and when—*regardless of what distance the pistons may travel in the brake cylinders*—their pressures become 25 pounds, or a very little greater, the same pressure being contained in chamber *b*, there is an equalization of pressure on piston 10, and the spring reacting upon valve 5 closes it and pressure supply to the brake cylinder ceases—until the pressure of chamber *b* and the brake cylinders begins to reduce through leakage, whereupon the greater pressure in application cylinder G again unseats the application valve, 5, and the brake cylinders are resupplied up to equalization, when the valve is closed as before. This automatic pressure-maintenance will be continuous as long as the charge remains in the application chamber; but when an increase of brake-pipe pressure moves the equalizing valve to release position, the air in the application chamber is exhausted to the atmosphere in the same manner as was explained in reference to the triple valve; or its discharge may be effected by the independent brake-valve (through a pipe connection not shown in the Fig. 5 diagram); but in either or any case, when the pressure of the application chamber is reduced wholly or in part, an *exhaust valve* which is not shown automatically releases the pressure from the brake cylinders in conformity to the exhaust of the application-chamber air—*pound for pound;* such additional functions of the distributing valve will be explained in detail further along, as at this time it is only desired to illustrate the application principle.

THE NO. 6 E-T DISTRIBUTING VALVE.

The **distributing valve** as a whole, consists of two sections bolted firmly together (see Fig. 6), one of which contains the operating apparatus—valves, pistons, etc.—and may be subdivided into two portions, the lower, or "equalizing portion," which we have already compared to a triple valve, and the upper, or "application portion," that directly controls the flow of pressure from main reservoir to brake cylinders, and from brake cylinders to the atmosphere in releasing the brake (see Fig. 7); the other section is called the "double-chamber reservoir," and it is also divided, as the name implies, by a partition or bulkhead which is part of the main-body casting, and air-tight, into two chambers which are called the "pressure chamber," and the "application chamber" (Fig. 6), and which will be understood as corresponding to an *auxiliary reservoir* and a *dummy brake cylinder*, respectively, to furnish the pressure-volumes for the correct operation of the equalizing, or triple valve, portion of the distributing valve; and the application chamber is ordinarily in connection with the *application cylinder* (in Fig. 7, the space closed by the cylinder head, 7), in which its pressure acts upon *application piston* 10, in part as described in connection with the diagrammatic Fig. 5.

No. 6 Distributing Valve

In the ideal sketch of Fig. 5, the pressure chamber and application chamber were shown in comparatively the same size as the regular auxiliary reservoir and brake cylinder of the automatic-brake system, and it was explained that the ratio of pressure supply to the application chamber and application cylinder conforms to the normal pressures obtained in brake cylinders of the common automatic system; as long as the sizes of an auxiliary reservoir and brake cylinder are *proportionately the same* their actual sizes may be reduced or increased to any extent without changing the ratio of equalization of pressure between them; and as the sole duty of their E-T counterparts are to furnish pressure to the comparatively small application cylinder, the reservoir containing the pressure chamber and application chamber is made so small as to take up but little room.

Referring to Fig. 7: It should be understood at first that the equalizing portion and pressure chamber are used in automatic applications only, service reductions of brake-pipe pressure causing the equalizing valve to connect the pressure chamber to the application chamber and application cylinder, allowing air to flow from the former to the latter two—to the application chamber, for expansion to the pressure equivalent to that which is desired in the brake cylinders, and to the application cylinder as the actuating power to be applied to the *application side* of piston 10 (upper portion). The upper slide valve, 5, connected to the spindle, or stem, of piston 10, holds main-reservoir pressure above it and admits a graduated amount of it to the brake cylinders when the locomotive brake is applied—an amount to correspond to the pressure in the application cylinder—and is called the "application valve," while the under one, 16, is used to release the pressure from the brake cylinders and is named the "exhaust valve"; in Fig. 7 the space between piston 10 and the head, 7, is the application cylinder, and the whole space to the right of piston 10 as far as cap-nut 22 is in permanent connection with the locomotive-brake cylinders; any greater pressure in the application cylinder than may be in the brake cylinders will, it can be plainly seen, force the application piston, 10, to the right, to close the exhaust valve and open the application valve, admitting main-reservoir air to the brake cylinders until their pressure equals that in the application cylinder; also, any variation of application-cylinder pressure will be exactly duplicated in the locomotive-brake cylinders, and the resulting pressure maintained regardless of almost any brake-cylinder leakage.

It is obvious that the pressure supply to the brake cylinders of the engine and tender is thus practically unlimited, but the limit has been found in some few cases when the brake-cylinder piston-packing-leathers have been partially blown out, or one of the brake-cylinder pressure-supply pipes has become broken off, and the braking pressure has escaped faster than it could be resupplied through the very large port of the application valve, or faster than the pump could com-

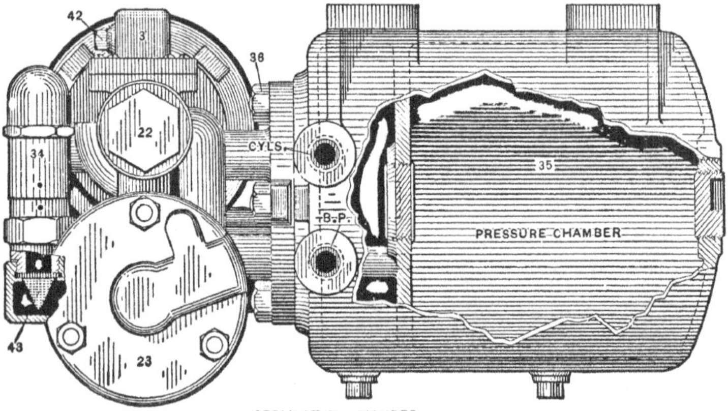

FIG. 6.—No. 6 Distributing Valve and Double-chamber Reservoir. CYLS—brake cylinder pipe; BP—brake pipe.

press it. The whole operation of the locomotive brake, therefore, consists in admitting and releasing air pressure into or out of the application cylinder; in independent applications, directly through the independent brake-valve; in automatic applications, by means of the equalizing portion and the air stored in the pressure chamber.

The well-known principle embodied in the Westinghouse quick-action triple valve, by which it gives a high brake power in emergency applications, and a sufficiently lower one, in full-service applications, to provide a desired protection against wheel sliding, is embodied in the No. 6 distributing valve, but without the violent shock to the brake rigging from cylinder piston to brake shoes that occurs at an emergency application of the quick-action triple valve, and the venting of brake-pipe air is not included as an emergency feature unless specially demanded as an adjunct to the standard equipment; the emergency increase of application-cylinder pressure is accomplished by cutting off the application chamber from it, when the pressure chamber will equalize with the quite small application cylinder at a greatly increased pressure that will be followed by a correspondingly high brake-cylinder pressure.

NAMES OF THE OPERATING PARTS, NO. 6 DISTRIBUTING VALVE.

All of the operating parts are plainly shown in Fig. 7; and the faces and seats, and plan views of the equalizing slide valve and its graduating valve, are shown in Fig. 8. In connection with a study of Figs. 2 A, 2 B, and Fig. 6, the piping connections of Fig. 7 and the connecting ports between the reservoir section and the valve section will be readily understood. The SAFETY VALVE is an essential part of the distributing valve that will be described in detail further along. Referring to Figs. 6 and 7, the names of parts of this apparatus are as follows:

2, Body.
3, Application-Valve Cover.
4, Cover Screw.
5, Application Valve.
6, Application-Valve Spring.
7, Application-Cylinder Cover.
8, Cylinder-Cover Bolt and Nut.
9, Cylinder-Cover Gasket.
10, Application Piston.
11, Piston Follower.
12, Packing-Leather Expander.
13, Packing Leather.
14, Application-Piston Nut.
15, Application-Piston Packing-Ring.
16, Exhaust Valve.
17, Exhaust-Valve Spring.
18, Application-Valve Pin.
19, Application-Piston Graduating Stem.
20, Application-Piston Graduating Spring.
21, Graduating-Stem Nut.
22, Upper-Cap Nut.
23, Equalizing-Cylinder Cap.
24, Cylinder-Cap Bolt and Nut.
25, Cylinder-Cap Gasket.
26, Equalizing Piston.
27, Equalizing-Piston Packing-Ring.
28, Graduating Valve.
29, Graduating-Valve Spring.
31, Equalizing Valve.
32, Equalizing-Valve Spring.
33, Lower-Cap Nut.
34, Safety Valve.
35, Double-Chamber Reservoir.
36, Reservoir Stud and Nut.
37, Reservoir Drain-Plug.
38, Distributing-Valve Drain-Cock.
39, Application-Valve-Cover Gasket.
40, Application-Piston Cotter.
41, Distributing-Valve Gasket (not shown).
42, Oil Plug.
43, Safety-Valve Air Strainer.
44, Equalizing-Piston Graduating Sleeve (numbered 60 on all subsequent plates).
45, Equalizing-Piston Graduating-Spring Nut.
46, Equalizing-Piston Graduating Spring.

Interior of Distributing Valve

To simplify the tracing of the ports and connections, the various positions of this valve are illustrated in nine *diagrammatic* views; that is, the valve is distorted

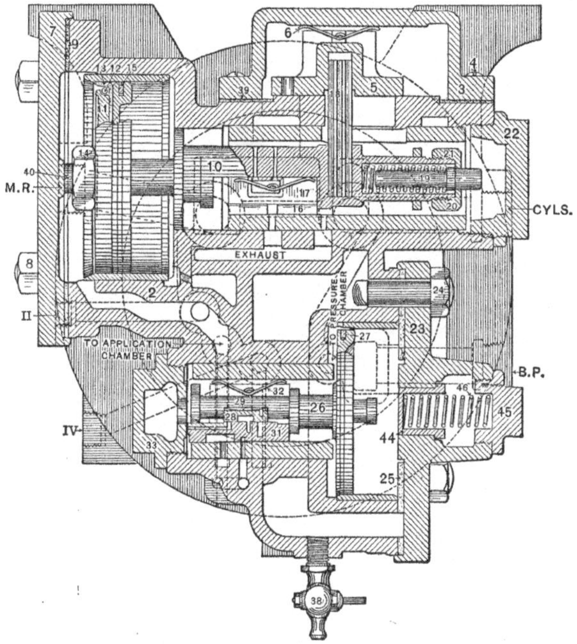

Fig. 7.—No. 6 Distributing Valve. Connections: MR—main-reservoir pipe; IV—distributing-valve release pipe; II—application-cylinder pipe; CYLS—brake-cylinder pipe; BP—brake pipe.

FIG. 8. — Graduating Valve, Equalizing Valve, and Equalizing-valve seat of No. 6 Distributing Valve.

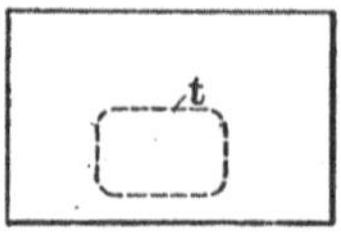

PLAN OF GRADUATING VALVE

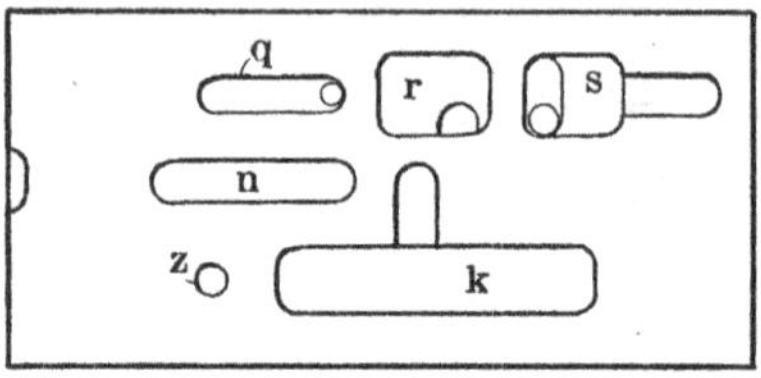

FACE OF SLIDE VALVE

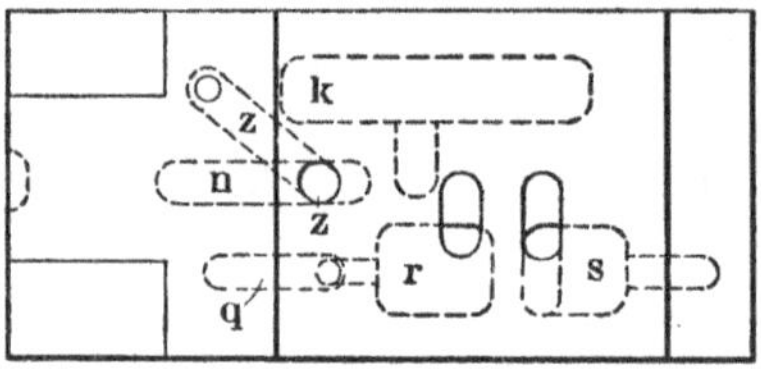

PLAN OF SLIDE VALVE

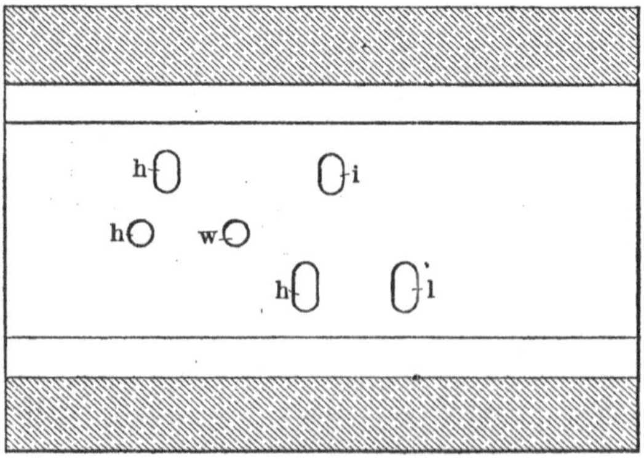

PLAN OF SLIDE VALVE SEAT

to show the parts differently than actually constructed with the object of explaining the operation clearly, instead of showing exactly how they are designed. The chambers of the reservoir are for convenience indicated at the bottom as a portion of the valve itself. In Fig. 7, equalizing piston 26, graduating valve 28, and equalizing slide valve 31 are shown as actually constructed; but as there are ports in the valve that can not be indicated at once in sectional side elevation just as they exist (see Fig. 8—*face of slide valve*, and *plan of seat*), the diagrammatic illustrations show each slide valve considerably elongated so as to make all the ports appear in one plane, with similar treatment of the equalizing-valve seat. Fig. 8 shows the correct location of these ports.

Referring to Fig. 7, the port through application valve 5 is of greater area than appears in sectional side view, as it extends transversely to nearly the width of the valve, and in full application position is in register with a port exactly corresponding in plan and area in the seat; from the center of the latter port a narrow, longitudinal opening is cut through the valve seat, but always covered by the valve, for the traverse of application-valve pin 18.

A piping diagram accompanies each chart in the following series that represents the distributing valve in the different operating positions, showing the contained pressures as they are affected, primarily, by the operation of the engineer's brake-valves, and secondarily, by the action of the distributing valve; the only omissions of the piping diagram being where it would merely be a repetition, as in connection with the charts showing the distributing valve when it has automatically returned to *lap position* after an application, etc. A COLOR KEY is supplied for reference on each color-plate page.

EXPLANATION OF THE DIAGRAMMATIC CHARTS OF THE NO. 6 E-T EQUIPMENT.

Running Position.

Figures 9 A and 9 B: Fig. 9 A shows the No. 6 E-T EQUIPMENT as a whole, with the automatic and independent brake-valves in **running positions,** and pipes and reservoirs charged with pressures as indicated by their colors; Fig. 9 B is a diagrammatic chart of the distributing valve with the operating parts in **release** and **charging position,** as the result of maintenance of brake-pipe pressure and the absence of pressure in the application cylinder.

Referring to Fig. 9 A: The boiler-pressure steam (dark blue) enters the steam cylinder of the air pump, enforcing action of the steam piston and the connected air piston in the air cylinder, and is exhausted through the pipe (light blue), that leads to the smoke box, or main exhaust passages of the locomotive cylinders. Atmospheric air (orange) is drawn in through the strainer to the air cylinder, in which it is condensed to main-reservoir pressure (red), and passes through both main reservoirs directly to the automatic brake-valve, the feed valve, the reducing valve, the high-pressure governor top, the large duplex gauge where it is indicated by the red hand, the by-pass strainer-and-check-valve, and the distributing valve, and indirectly, through the automatic brake-valve to the excess-pressure governor top. The feed valve is adjusted to supply 70 pounds of pressure in all branches of service except the High-Speed Brake, to the feed-valve pipe (brown), which delivers this air to the automatic brake-valve, and through a small connecting pipe to the regulating-spring case of the excess-pressure governor top. With the automatic brake-valve in **running position,** as shown, or in **holding position,** the feed-valve pipe air passes through the rotary valve to a branch of the brake pipe (yellow), at the same pressure, and from which there are two connecting pipes, one leading to the small duplex gauge on which the brake-pipe pressure is registered by the black hand, and the other one to the cut-out cock in the by-pass arrangement used in charging the air equipment of a "dead" engine, this branch pipe teeing into the main brake pipe that leads to all triple valves of the train, and from which the air passes through the large strainer-tee, and branch pipe, to the distributing valve. The reducing valve regulates the air supply to the reducing-valve pipe (lavender), at 45-pounds pressure which feeds to the independent brake-valve, and has a branch connecting to the signal line strainer-and-check; beyond this fitting the same amount of pressure ensues, but it is then called the signal-line pressure (purple). The same pressure that

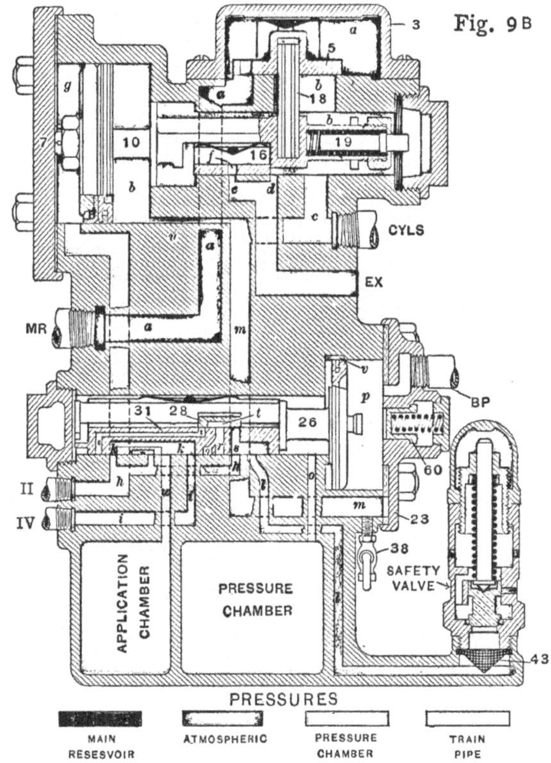

Fig. 9 b.—No. 6 Distributing Valve in Released and Charging Position.

Fic

Fig. 9A

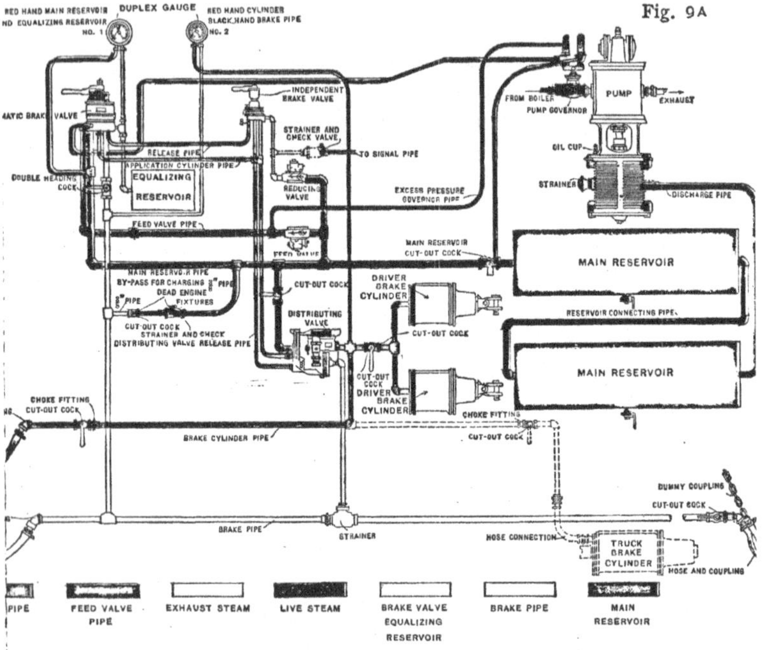

am, No. 6 E T Equipment. Colors showing sequence of pressures, with the automatic an independent brake-valves in running position.

Fig. 9B

Released

Fig. 9A

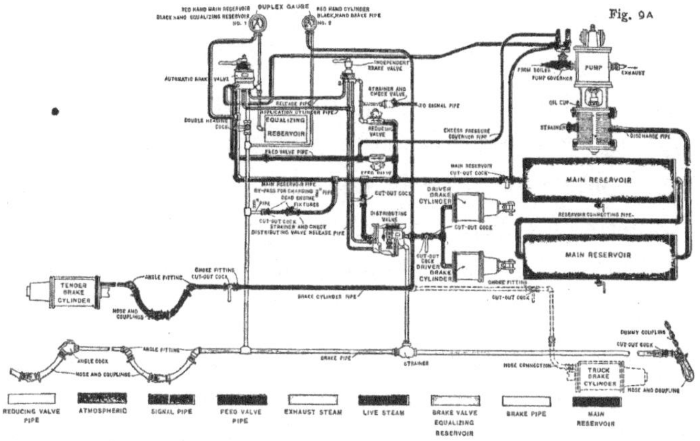

Fig. 9 A.—Piping Diagram, No. 6 E T Equipment. Colors showing sequence of pressures, with the automatic and independent brake-valves in running position.

is passing to the brake pipe—70 pounds—is also supplied through the rotary valve of the automatic brake-valve to the pipe leading to the equalizing reservoir (green), and its branch to the large duplex gauge, whereon its pressure is registered by the black hand (sometimes called "chamber D pressure," because it is contained in that chamber of the automatic brake-valve).

With both brake-valves in **running position,** all the rest of the pipes of the equipment are open to the atmosphere, or at least not containing pressure any greater than atmospheric (as in that portion of the "by-pass attachment" where it is separated from the brake-pipe pressure by the closed cut-out cock, and the check-valve prevents main-reservoir air from entering), and are given the atmospheric color (orange).

Fig. 9 B represents the distributing valve as near like as possible to the sectional view in Fig. 7, and such arbitrary changes in the location of parts and ports as have been made in the interest of a clear understanding have been explained; piston 10 and attachments represent the upper, or *application,* portion; piston 26 the lower, or *equalizing,* portion—a triple valve, in effect—while the pressure chamber and application chamber together form the reservoir section, shown as a dark, circular background in Fig. 7. Unreduced main-reservoir pressure enters at MR and fills chamber *a;*

brake-pipe pressure is represented as entering through the cylinder cap, 23, and is contained in chamber *p;* the pipe connecting at CYLS is the locomotive brake-cylinder pipe; EX is the brake-cylinder exhaust port; II is the connection of the application-cylinder pipe, and IV that of the distributing-valve release pipe.

All parts are in the **release** and **charging** (or **charged**) **positions:** it is not necessary that the distributing valve shall be in a charged condition for the parts to be in this position, as there is nothing to cause them to change their locations if the air pump should be shut off and the pressures die down; release position of the lower portion will be taken as the result of brake-pipe recharge; of the upper portion, when the automatic and independent brake-valves are both in **running position** and the equalizing portion of the distributing valve in **releasing position,** or with the independent brake-valve in **release position** *under any circumstances.*

Brake-pipe pressure in chamber *p*, having forced equalizing piston 26 to the extreme left, finds a passage past the piston through feed-groove *u* into the compartment surrounding the slide valves, and the pressure chamber (green); and the piston has so placed the graduating valve, 28, that the ports *q*, *r*, and *z*, and cavity *t* are blanked against all other communication; through the ports in equalizing slide valve 31 and the slide-valve seat, application cylinder *g*, the application chamber, and the safety valve, are all brought into open communication with each other—as plainly shown by the arrangement, and sequence of color—and also with the application-cylinder pipe which is blanked at the rotary valves of the automatic and independent brake-valves, and with the distributing-valve release-pipe which being routed through both brake valves finds an opening to the atmosphere at the large exhaust port of the automatic brake-valve; hence it is that the greater space of the distributing valve is shown in the subject plates to contain only atmospheric pressure (orange), for with application cylinder *g* emptied of actuating pressure, any remaining pressure above atmospheric in chamber *b* would place application piston 10 in the **release position** as shown, in which the brake-cylinder pressure would escape past the end of the exhaust valve 16 and through port *f* in that valve, to ports *e* and *d* in the seat, and to the atmosphere at EX, and as chamber *b* is always in direct communication with the brake cylinders the released condition is complete: application slide valve 5 being fixed in its closed position by the engagement of pin 18 which is fitted neat in a socket in a spindle of the application piston.

The edge of piston 26 is made practically air-pressure-tight by a metallic packing ring, same as in an ordinary triple valve; and the application piston, 10, also is fitted with a similar packing ring, but, as it is extremely important that the latter piston shall be as nearly leakage-proof as possible, it also carries a *packing leather* of the same style as the packing of the brake-cylinder pistons, with the usual expanding ring within it for keeping the bearing surfaces of the leather in permanent contact with the walls of the application cylinder.

It is a common impression that the drain-cock, 38, is for the purpose of draining off the moisture, etc., from the equalizing portion of the distributing valve, because it is located just beneath the lower portion, but it will be seen that it is to drain chamber *b* and that portion of the cylinder containing piston 10, on the right, the large passage, *m*, trapping the moisture that is brought in with the main-reservoir air before it can pass on to the locomotive-brake cylinders, and permitting its removal through the drain-cock; if for any reason it should ever become necessary to bleed the locomotive-brake cylinders, it is apparent that this can be done by opening drain-cock 38; and if it should be left open through accident, under ordinary circumstances, it would have the effect of a bad leak of brake-cylinder pressure that would keep application valve 5 partly open all the time during a brake application, and represent an undesirable waste of main-reservoir air.

It is hardly necessary to explain that the top side of the large equalizing slide valve, 31, is faced off to form the seat for the small graduating-valve, 28; and that the latter valve is so closely connected to the equalizing piston as to be, in movement, a part of it, while there is enough slack between the ends of the equalizing valve and the shoulders of the piston spindle to permit of a short independence of movement of the piston, this being identically the same as in the instruction design of a triple valve, Figs. 4 A to 4 D, inclusive.

The spindle of application piston 10 has a tubular end containing the application piston graduating-stem, 19, confined between the graduating-stem nut and the graduating spring, and the duty of these parts is to assist the application mechanism in taking the lap position as the termination of an application movement.

Feed-groove *u* is of a size that permits the pressure chamber to be charged from the brake pipe in about the same time that is required for the auxiliary reservoirs of the cars to charge—approximately one pound of pressure-chamber increase per second.

It will be considered, in connection with the following charts of the distributing valve and piping diagrams, that the feed valve is adjusted to supply 70 pounds pressure to the brake pipe; that the excess-pressure governor top regulates the pump at 90 pounds in the main reservoir while the automatic brake-valve is in **release, running, or holding, positions**; that the high-pressure governor top has control of the pump when the brake-valve is in positions other than stated, and will permit of 110 pounds as the maximum main-reservoir pressure.

Automatic Service-Application Position.

Figs. 10 A and 10 B represent a service application by the automatic brake-valve—say a 10-pound brake-pipe reduction (refer to Fig. 10 A); this is primarily made by a 10-pound reduction of the equalizing-reservoir pressure which is indicated by the black hand of the large duplex gauge, and this *automatically* causes an equal reduction of brake-pipe pressure through the equalizing-discharge valve of the brake-valve, the latter pressure being indicated by the black hand of the small duplex gauge.

The broken colors on the piping diagram indicate these reduced pressure conditions, and also applies to pressures below the normal; the small pipe that carries full main-reservoir pressure to the excess-pressure governor top when the automatic brake-valve is in **running position** is cut out from that supply when the brake-valve is moved toward the **application position** —as now—and its pressure is *minus;* the low-pressure governor top is thus temporarily inactive, and the pump starts up to add to the 90-pounds excess pressure in the main reservoir until it reaches 110 pounds when the action of the high-pressure governor top will stop it at that figure.

Fig. 10^B

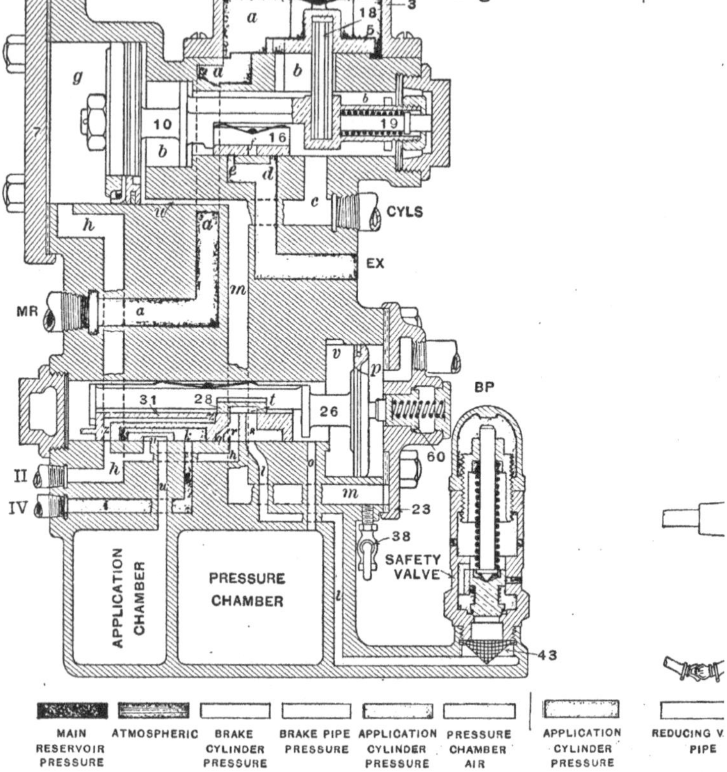

Fig. 10 b.—No. 6 Distributing Valve in Automatic Service-Application Position.

Fig. 10A

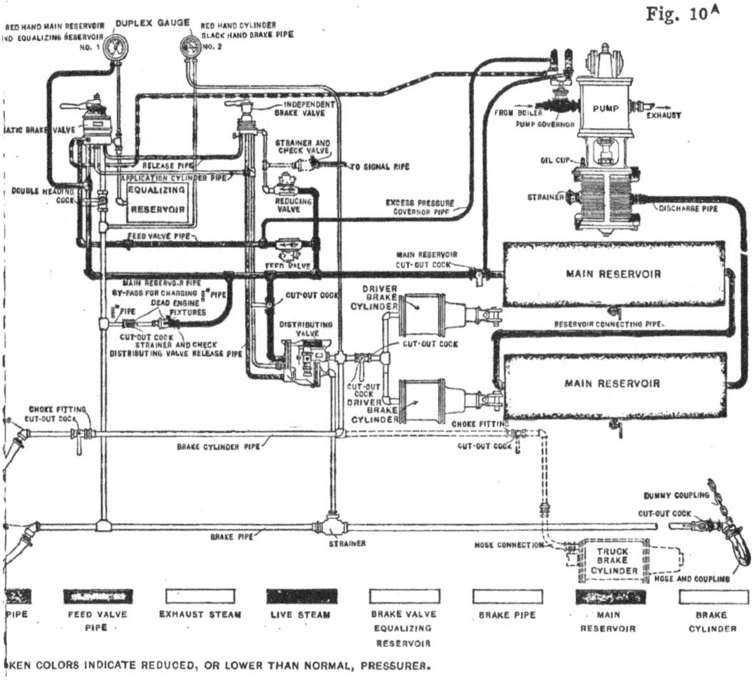

am of No. 6 E T Equipment. Colors showing sequence of pressures, with the automatic brake- service-application position; independent brake-valve in running position.

g. 12B

AKE
INDER
SSURE

gency

Fig. 12A

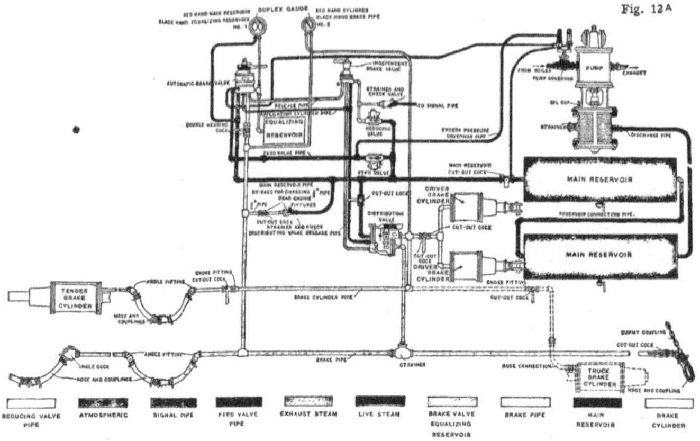

Fig. 12 A.—Piping Diagram of No. 6 E T Equipment. Colors showing sequence of pressures, with automatic brake-valve in emergency-application position, and independent brake-valve in running position.

Automatic Service Application

This reduction of brake-pipe pressure induces an actuating pressure of about 25 pounds in the application cylinder, which flows into the application-cylinder pipe and to both brake-valves, where the outlets are blanked; the distributing-valve release pipe is no longer in connection with the atmosphere, its outlet at the automatic brake-valve having been closed by the application movement of the rotary valve, and it is also blanked against pressure at the distributing valve, and so retains the atmospheric reference color of *orange.*

The distributing valve now causes main-reservoir air to flow to the several brake cylinders of the locomotive until their pressure equals that of the application cylinder, the brake-cylinder pressure registering by the red hand of the small duplex gauge.

Having learned the general pressure movements that induce action of the distributing valve we will now refer specifically to that unit of the E-T equipment which automatically graduates and maintains the locomotive braking pressure with such wonderful nicety—the Distributing Valve, Fig. 10 B: The 10-pound reduction of brake-pipe pressure in chamber *p* has permitted the 70 pounds in the pressure chamber to move equalizing piston 26 to the right until its button head strikes the graduating sleeve, 60, in which position it stops, as shown, without completing its maximum travel for two reasons: first, the resistance of the graduating spring against the graduating sleeve; second, in that position the pressure on the left of piston 26 begins to reduce about as rapidly as the brake-pipe pressure is being discharged, for the movement of the piston has pulled graduating valve 28 to open port *z* in the big slide valve, 31, and the latter—the equalizing valve—has been drawn by a shoulder of the piston spindle to a position in which port *z* is in register with port *h* in the seat through which pressure-chamber air flows to application cylinder *g* and, in connection with cavity *n* in the face of the equalizing valve and port *w* in its seat, to the application chamber whose only purpose is to supply the space necessary for the proper expansion of this actuating and regulating pressure.

Proportionate to the amount of brake-pipe reduction we have 25 pounds pressure in cylinder *g*, and this has forced application piston 10 the full length of its right-hand stroke, compressing the spring in its tubular end when stem 19 strikes the cap nut, and by means of the engagement of pin 18 with the application valve, 5, moving that valve to the right until the large port through it is in even register with the port in the seat; a shoulder of piston 10 has engaged with the exhaust valve, 16, sliding it to the right and closing exhaust ports *e* and *d* in its seat; main-reservoir air now flows from chamber *a* through the port in valve 5 to chamber

Fig. 11

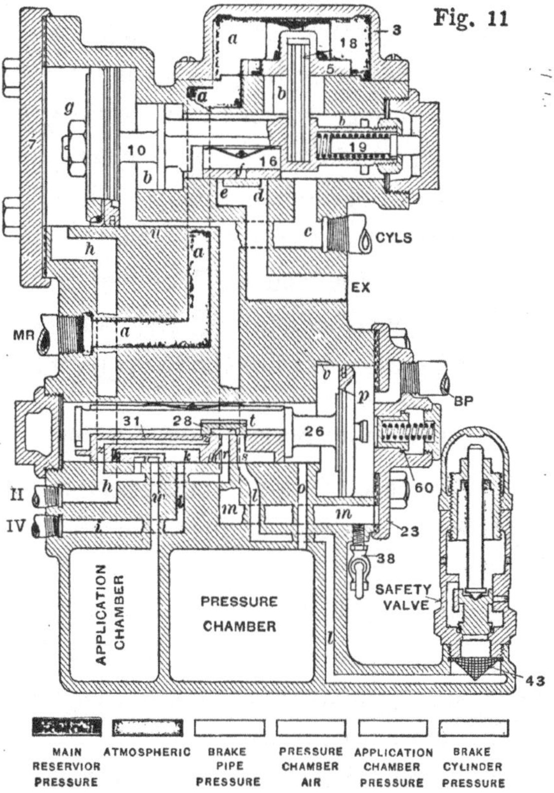

Fig. 11.—No. 6 Distributing Valve in Automatic Service Lap Position.

b, and thence through port *c* and the pipe connecting at CYLS to the brake cylinders.

In this position the application-cylinder air has its fullest connection with the safety valve (so arranged specially for High-Speed Brake service), cavity *t* in the graduating valve, connecting ports *r* and *s* in the equalizing valve, through which the pressure from port *h* flows to passage *l* and the safety valve, as shown.

The sequence of pressures in the pipes connecting with the distributing valve at II and IV has been explained in Fig. 10 A, and the blanking of the distributing-valve connection of the latter port is here seen in port *i* and the dead cavity *k* in the equalizing valve.

Automatic Service-Lap Position.

As the sequence of an automatic-service application the parts within the distributing valve assume the position of **service lap,** as represented in Fig. 11; this phase involves no change in the pressures as indicated in the piping diagram that precedes this chart, and it will only be necessary to refer to certain changes incident to the distributing valve itself.

It was assumed that the brake-pipe pressure had been reduced from 70 to 60 pounds; and when the pressure on the left of equalizing piston 26 had become a trifle less than the latter figure, from loss of air to the application chamber and application cylinder, the pressure in chamber *p* being somewhat the stronger pushed piston 26 to the left until the right-hand shoulder of its stem striking equalizing valve 31 the piston was stopped in the position shown; equalizing valve 31 was not moved, and the short, *back lash* of the piston effected nothing directly except to move graduating valve 28 "on lap," the latter valve closing port *z* to cut off further increase of application-cylinder pressure, and closing communication between the application cylinder and the safety valve by cutting off port *r* from port *s* in the equalizing valve. (It is unnecessary for the safety valve to be cut in to the application-cylinder pressure when there is no chance for that pressure to increase, as in the lap position; and if the safety valve were not cut out at this time, an obstruction between the valve and valve-seat would result in complete loss of the application-cylinder pressure and consequent release of the locomotive brakes.)

Referring to the upper portion of the figure: In the preceding chart the pressures were increasing on both sides of the application piston, 10, and 25 pounds had accumulated in application cylinder *g* when its further increase was stopped by the closing of the graduating valve (lower portion); when that occurred, main-reservoir air from chamber *a* continued to flow to chamber *b* and the locomotive brake cylinders until the pressure in the latter was probably a little greater than that in cylinder *g*, when the reaction of the coil spring in the end of the spindle of piston 10 moved the piston and application valve 5 to the left, cutting off the communication between chambers *a* and *b*, as shown, and which temporarily closes off pressure supply to the brake cylinders. The leftward movement of piston 10 ceased just as the application valve came to its lapped position, partly because the frictional resistance of exhaust valve 16 under the air pressure of chamber *b* was met with, and at that point the extending force of the coil spring was checked by the shoulder of graduating stem 19 striking the graduating-stem nut.

I said that the supply of air to the brake cylinders was *temporarily* closed off, because any subsequent leakage of the locomotive brake-cylinder pressure will permit the application piston to move again to the right to application position, and the brake cylinders to be resupplied at the original pressure; or, if any further reduction of brake-pipe pressure shall be made, both the upper and lower portions of the distributing valve will assume the positions shown in the service-application chart, following with the **service-lap position** in which the only difference from the subject chart will be suggestive—lower brake-pipe and pressure-chamber pressures, and higher application-cylinder and brake-cylinder pressures.

When the brake-pipe pressure has been reduced

by 20 pounds, the contents of the pressure chamber will equalize fully with the application-chamber and application-cylinder at a pressure of 50 pounds; any further reduction will be of no effect on the locomotive brakes, although a 25-pound reduction is necessary to insure the full-on application of all car brakes; further than that, any reduction of the brake-pipe pressure is a waste of air, and in such cases the lower portion of the distributing valve will take the same position as will be shown in Fig. 12 B, **emergency position,** next to follow.

Emergency-Application Position.

Figs. 12 A and 12 B represent an emergency application as the result of placing the automatic brake-valve in **emergency position,** although the distributing valve would assume identically the same position from a service brake-pipe reduction of 25 pounds, or more —with the exception that in the latter case the safety valve would probably not be unseated.

Referring first to the piping diagram, Fig. 12 A, the movement of the automatic brake-valve to **emergency position** has effected a quick and heavy reduction of the brake-pipe pressure, as indicated by the broken color lines, and the effect on the distributing valve is to cause it to supply main-reservoir air to the locomotive-brake cylinders; the supply of main-reservoir pressure to

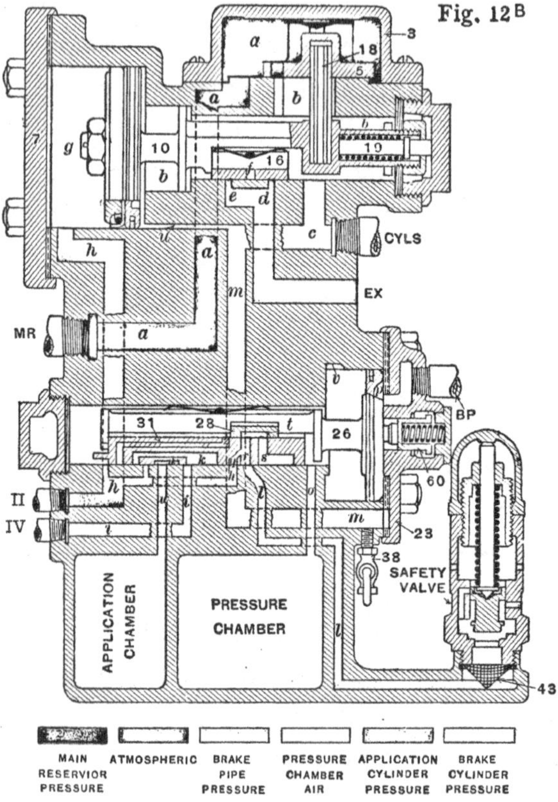

Fig. 12 b.—No. 6 Distributing Valve in Emergency Position.

Fig. 12A

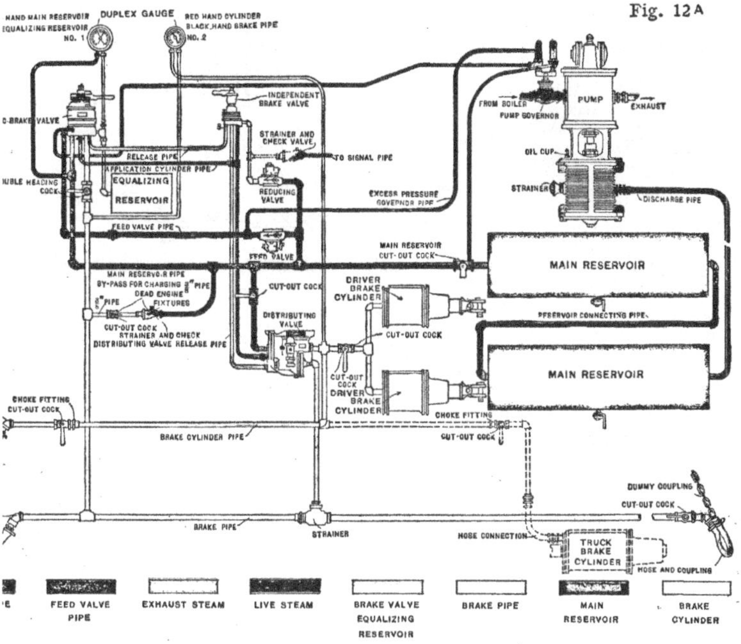

ram of No. 6 E T Equipment. Colors showing sequence of pressures, with automatic brake-
plication position, and independent brake-valve in running position.

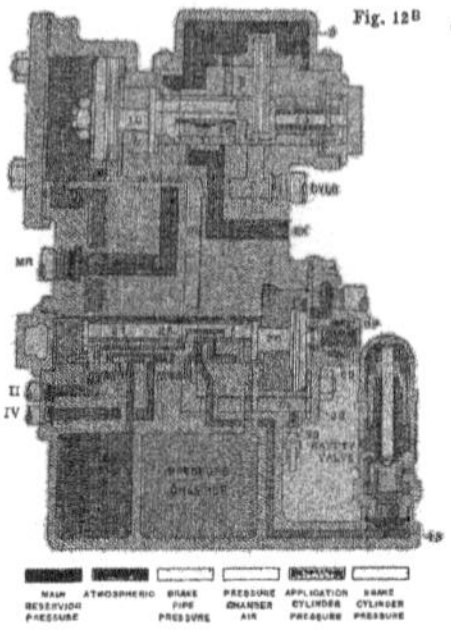

Fig. 12 B.—No. 6 Distributing Valve in Emergency Position.

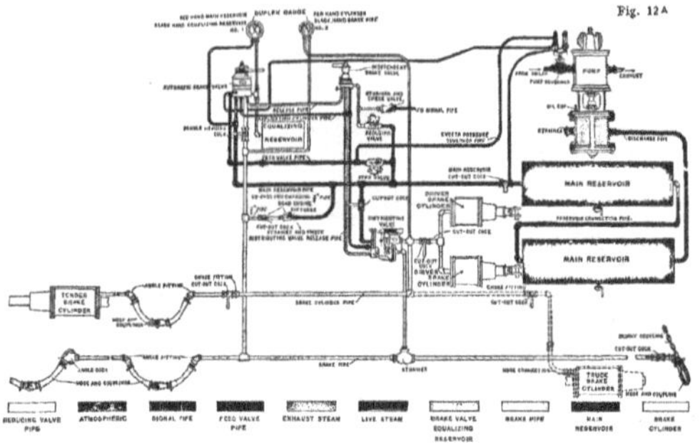

Fig. 12 A.—Piping Diagram of No. 6 E T Equipment. Colors showing sequence of pressures, with automatic brake-valve in emergency-application position, and independent brake-valve in running position.

the excess-pressure governor top has been cut out by the turning movement of the rotary valve of the automatic brake-valve, and the pressure remaining in the pipe leading from brake-valve to governor is slowly reducing, also apparent from its broken color line, thus cutting out the low-pressure governing effect, and the pump is now under control of the high-pressure governor top, only; the equalizing-reservoir air has no part to play in an emergency application, but its pressure is being slowly discharged, as indicated, through the automatic brake-valve; the distributing-valve release pipe is blanked at each terminal—the automatic brake-valve, and the distributing valve—and still retains atmospheric pressure; while the application-cylinder pipe is engaged in supplying main-reservoir air, as the *maintaining pressure,* from the automatic brake-valve to the application cylinder of the distributing valve.

Turning our attention now to the distributing valve, individually, the lower portion particularly (Fig. 12 B), equalizing piston 26 has been impelled to the right by the reduction of brake-pipe pressure somewhat the same as explained in connection with Fig. 10 B, except that in this case the heavier and sudden reduction permitted a stroke so rapid that the graduating spring was compressed and the piston completed its traverse, seating against the cylinder-cap gasket; the pressure-chamber air is flowing to the application cylinder, but into port *h* past the end of the equalizing slide valve instead of through its graduating port, *z*, the lower end of which is blanked on the equalizing-valve seat; it will be noticed that piston 26 has carried equalizing valve 31 to a position in which it has cut out the application chamber from its association with the application cylinder, and this so reduces the space in connection with the latter that the pressure-chamber air equalizes with the application cylinder at 65 pounds—instead of the 50 pounds obtainable at full-service applications.

Through ports *q* and *r* in the equalizing valve, the application-cylinder pressure in port *h* is permitted to flow to port- and passage-*l* and the safety valve; but the safety valve is adjusted at 68 pounds, and it would not unseat if this emergency application had been initiated by a break-apart of the air hose, or by the brake-valve on another engine, or in any manner whatever except by emergency position of the automatic brake-valve on the same engine; in the latter and present case, main-reservoir pressure is admitted to the application cylinder, as previously explained, to increase the pressuretin the latter to a regulated extent, and as positive insurance against loss of application-cylinder pressure through leakage; in the present case it raises the latter pressure to 68 pounds, and the capacity of the safety valve is sufficient to prevent it raising any higher.

Safety Valve. High-Speed Service

Except in making a quicker stroke, the action of the application portion of the distributing valve at emergency is precisely the same as at any service application, which was fully described in reference to Fig. 10 B; the pressure obtained in chamber *b* and the locomotive brake cylinders will be higher at emergency applications, of course, corresponding as usual to the pressure in the application cylinder.

In High-Speed Brake service, where the brake-pipe pressure is regularly carried at 110 pounds, and the main-reservoir pressure is 130 pounds or more, an emergency application raises the application-cylinder pressure to 93 pounds which the safety valve at once begins to blow down; but the passage between cavity *q* and port *r* in the equalizing valve is so small that the ebb of application-cylinder pressure is just enough faster than the supply through the brake-valve, that the safety valve will decrease it in practically the same time and manner as is done by the high-speed reducing valve, until it is down to about 75 pounds; the reason why it does not blow down to 68 pounds—the pressure at which the safety valve is set—is because the inflow of air through the brake valve at the high main-reservoir pressure carried in high-speed service is equal, at 75 pounds, to the outflow through the small opening to the safety valve. The higher brake-cylinder pressure at emergency application in High-Speed service, obtainable with the No. 6 than with the No. 5 E-T equipments, makes it particularly applicable to passenger engines that operate the Type L, high-speed, passenger triple valves on the cars.

It is practically impossible for the E-T brake to leak off if there is no leakage of application-cylinder pressure; the zone of such possible leakage includes the application-cylinder head, application-cylinder pipe, distributing-valve gasket, and remotely, the rotary valves of the automatic and independent brake-valves, and under certain conditions the distributing-valve release pipe; also the application piston, 10, leakage past its packing forming one of the worst troubles incidental to the E-T equipment, being concealed; when an emergency application is made by the local automatic brake-valve, however, the bad effects of leakage at this point are usually overcome by the maintaining pressure from the main reservoir *via* the automatic brake-valve.

Emergency-Lap Position.

In this reacting position there is no lap of the lower portion of the distributing valve, as the brake-pipe pressure is almost, or wholly, discharged; all movable parts remain as in Fig. 12 B, until the brake-cylinder pressure slightly exceeds the application-cylinder pressure, when the application piston and application valve

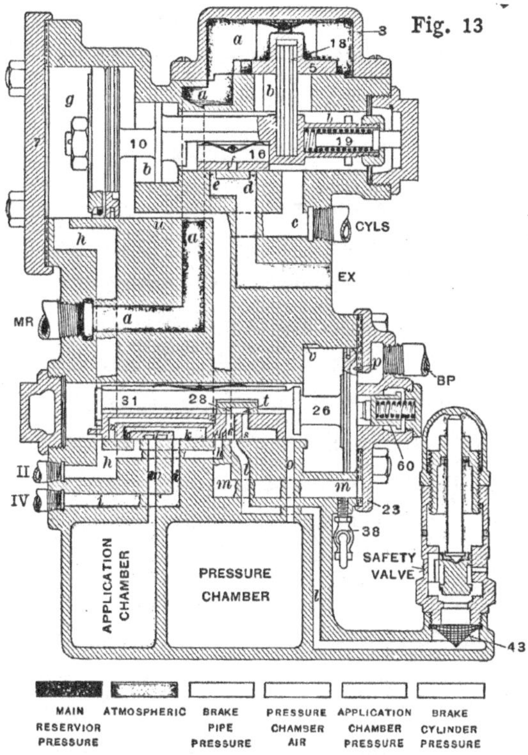

FIG. 13.—No. 6 Distributing Valve in Emergency Lap Position.

move back to the position known as emergency lap, as shown in Fig. 13; there is no change in pressures from the conditions indicated in Fig. 12 A, and that piping diagram applies as well in connection with this chart.

Automatic Release.

When the automatic brake-valve is placed in **release position** following a brake-pipe reduction by any manner of application, the recharge of the brake pipe will release all car brakes in the train, except those that may be held on by the cutting-in of their retaining valves, and the equalizing portion of the distributing valve on the locomotive will be moved to **release position,** without, however, releasing the locomotive brake, the application portion of the distributing valve remaining in the lapped position; see Fig. 14 B.

The benefit secured by this means is the ability of the engineer to release the brakes on the cars of a long freight train when the speed has been sufficiently reduced, yet to hold the locomotive brake applied so as to prevent the violent surging ahead of the forward portion of the train as the head brakes first release, and thus prevent the otherwise certain parting of the train.

Concerning the general pressure conditions incidental to this operation as indicated in the piping diagram,

Fig. 14 A, the automatic brake-valve having been placed in **release position** main-reservoir pressure finds the "open door" through that valve to the brake pipe, quickly increasing the pressure in the latter, and re-supplying the pipe to the diaphragm-valve section of the excess-pressure top of the pump governor. The increase of brake-pipe pressure will be registered by the black hand of the small duplex gauge, and flow to the triple valves throughout the train, releasing all car brakes, but to the distributing valve on the locomotive without releasing effect. Main-reservoir pressure is now flowing to the equalizing reservoir, and the register of the pressure in the latter by the black hand of the large duplex gauge can not be taken as a proper indication of brake-pipe pressure under present conditions (although before the advent of the No. 6 E-T brake this gauge hand was the only means of showing brake-pipe pressure). The feed-valve pipe now has no outlet, and contains its maximum charge of 70-pounds pressure. The application-cylinder pipe and distributing-valve release pipe are now in communication with each other, filled with the pressure of the application cylinder, and retained by reason of the outlet of the release pipe being blanked at the rotary of the automatic brake-valve; the result naturally being that the locomotive brake-cylinder pressure, of equal amount, is also retained as indicated in the coloring.

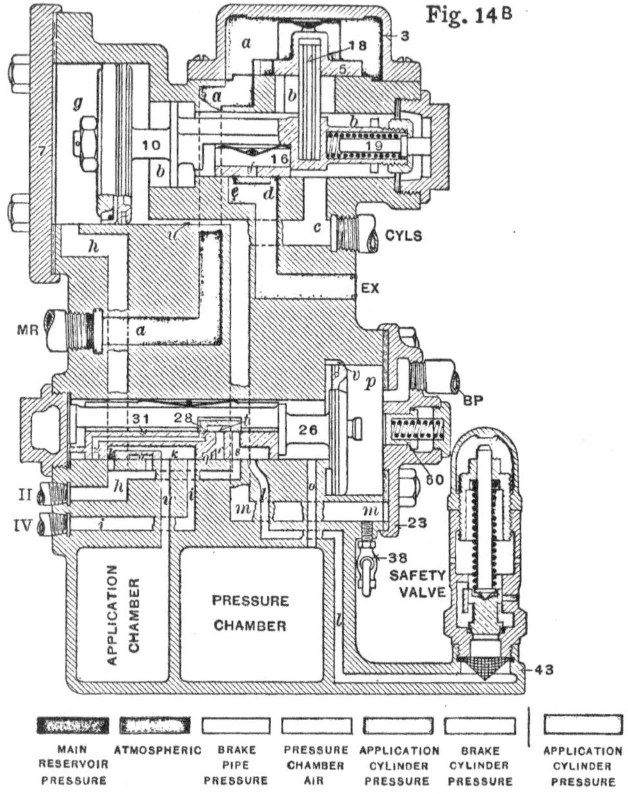

Fig. 14 b.—No. 6 Distributing Valve. Recharging position, with locomotive brake retained.

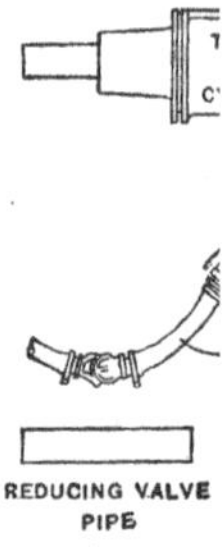

Fig. 1

Fig. 14 A

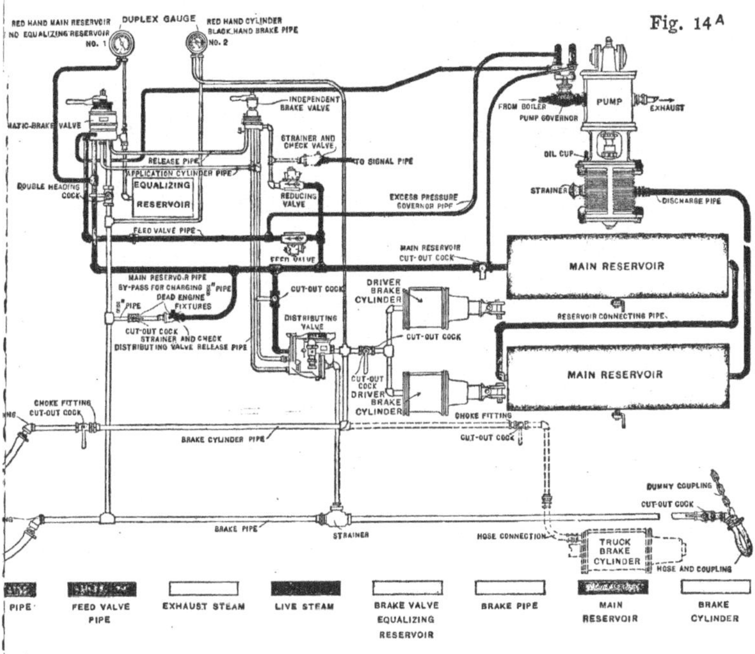

m No. 6 E T Equipment. Colors showing sequence of pressures at automatic-brake release. brake-valve in release position, independent brake-valve in running position.

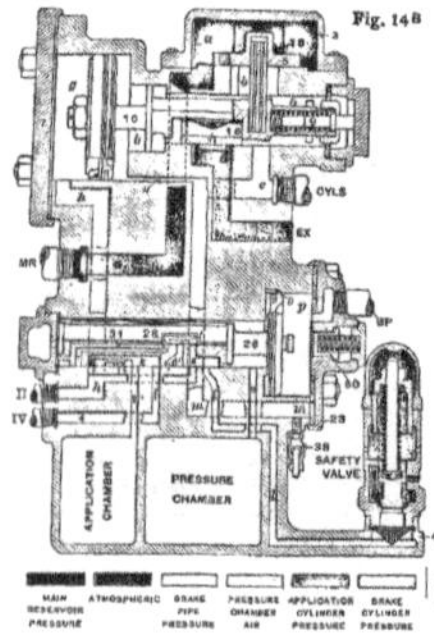

FIG. 14 B.—No. 6 Distributing Valve. Recharging position, with locomotive brake retained.

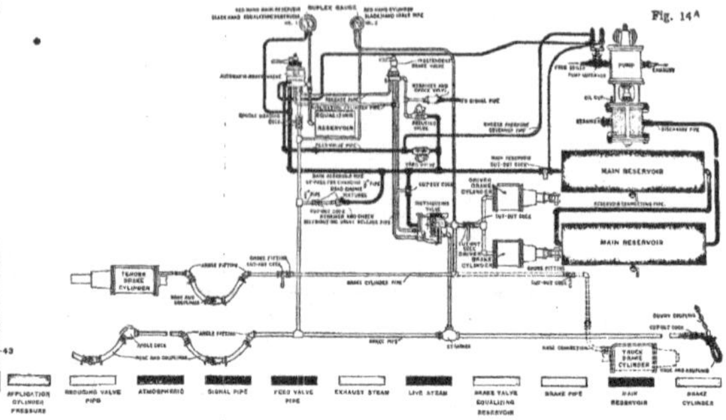

FIG. 14 A.—Piping Diagram No. 6 E T Equipment. Colors showing sequence of pressures at automatic-brake release. Automatic brake-valve in release position, independent brake-valve in running position.

In the DISTRIBUTING VALVE, FIG. 14 B, the increase of brake-pipe pressure in chamber *p* has forced the equalizing portion to release position, but as there can be no automatic release until the distributing-valve release pipe from IV shall be re-opened to the atmosphere by the automatic brake-valve being returned to **running position,** the only effect is to re-connect the application chamber with the application cylinder, and their combined volume with the distributing-valve release pipe. With no release of application-cylinder air, piston 10 and application valve 5, and exhaust valve 16, remain in the lapped condition as shown; there is no *maintaining pressure* being now supplied to the application cylinder, and the safety valve is seated.

If the **release position** of the automatic brake-valve had followed a service application, there would be no change made whatever in the distributing-valve pressures; but with the recharge of the brake pipe after an emergency application the application chamber is found empty of pressure, and its consequent equalization with the application cylinder when the equalizing valve is moved to **release position** somewhat reduces the pressure of the application cylinder, and this should be followed by a movement of application piston 10 to the left and a quick "pop" of brake-cylinder pressure escaping from the exhaust port at EX; the pressure in chamber *b* being thus relieved of its excess over that in application cylinder *g*, the application portion at once resumes the lap position as shown.

Meanwhile, of course, the pressure chamber is becoming recharged through feed-groove *u;* and before it (and the auxiliary reservoirs of the cars in the train) has had time to overcharge (above 70 pounds), the automatic brake-valve will be returned, either to **holding position** or **running position;** if the former, no change whatever occurs except that the brake pipe will receive its charge from the 70-pound feed-valve pipe; if the **running position** is taken, in addition to the brake-pipe supply being received from the feed-valve pipe the distributing-valve release pipe will be thrown open to the atmosphere at the automatic brake-valve and the locomotive brake will fully release; this phase being already fully described both in piping diagram and distributing-valve chart, Figs. 9 A and 9 B, "release and charging position."

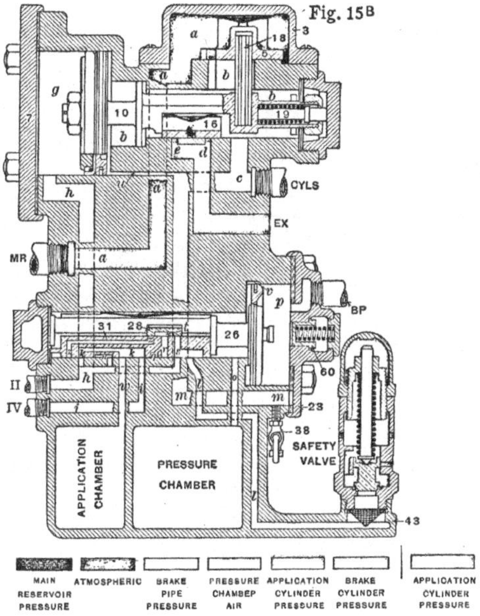

Fig. 15 b.—No. 6 Distributing Valve. Independent application position.

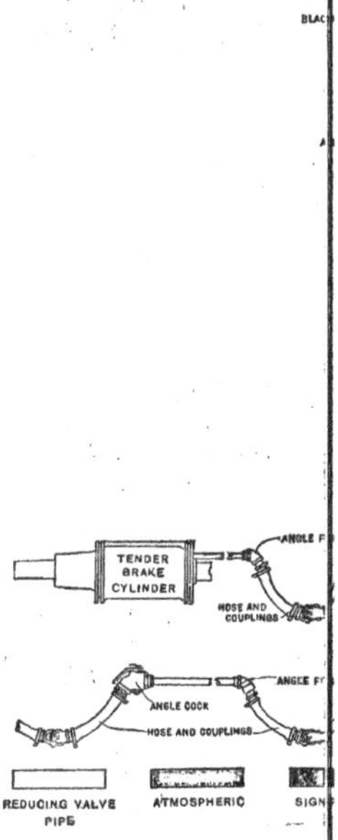

Fig. 15 a.—Piping Diag
valve in application p

Fig. 15A

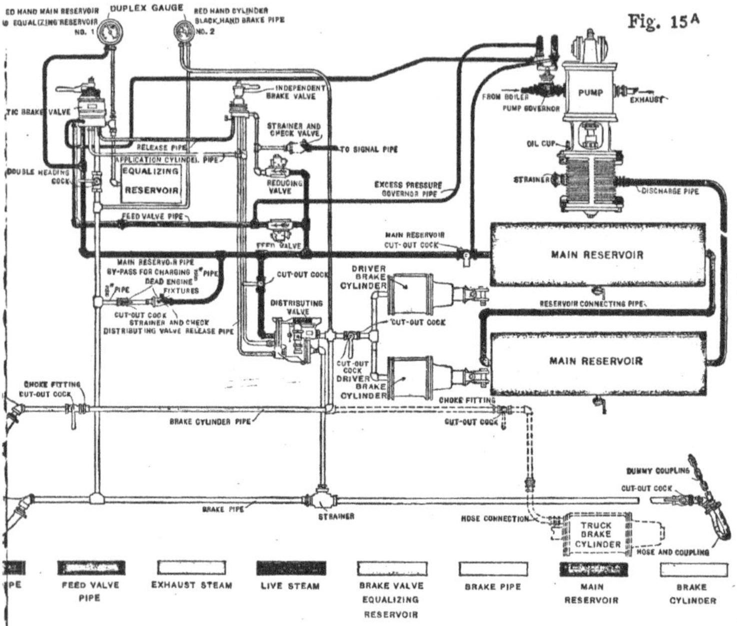

No. 6 E T Equipment. Independent locomotive-brake application. Independent brake-
ion, automatic brake-valve in running position. Colors showing sequence of pressures.

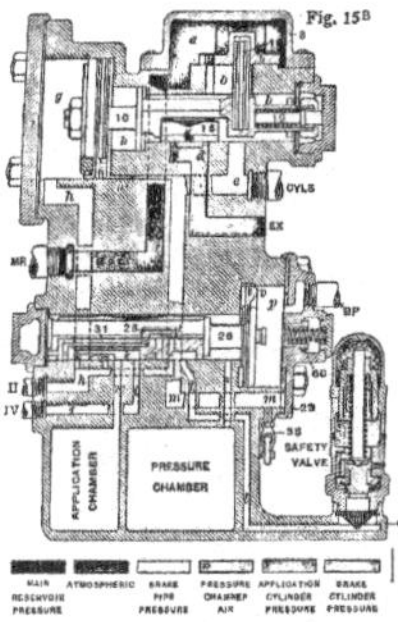

FIG. 15 B.—No. 6 Distributing Valve. Independent application position.

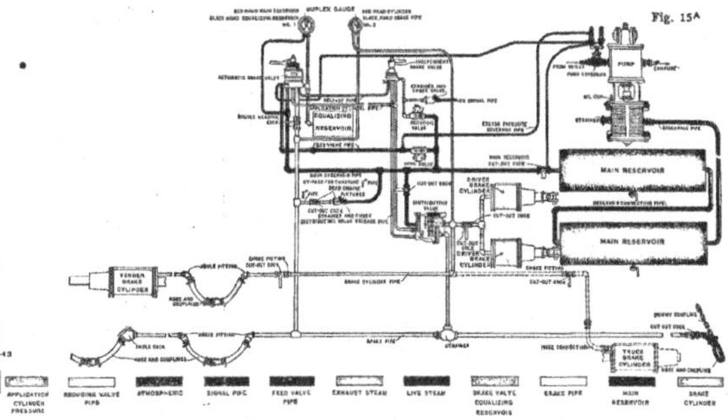

FIG. 15 A.—Piping Diagram No. 6 E T Equipment. Independent locomotive-brake application. Independent brake-valve in application position, automatic brake-valve in running position. Colors showing sequence of pressures.

INDEPENDENT LOCOMOTIVE-BRAKE APPLICATION,

BY INDEPENDENT BRAKE-VALVE.

Attention will now be given to the effects from the operation of the engineer's INDEPENDENT BRAKE-VALVE, first, in applying the locomotive brake while the automatic brake-valve remains in **running position,** and secondly, the consequent position of **independent lap** taken by the distributing valve; followed by a description of the manner in which the driver and tender brakes may be released independently of the train brakes after all have been automatically applied, and while the automatic brake-valve remains **on lap.**

The first condition is represented in Figs. 15 A and 15 B—**Independent Locomotive-Brake Application.** Referring to the piping diagram, the automatic brake-valve is in **running position,** and all pipes supplied by it with pressure are charged as explained in its common, **running position.** The independent brake-valve handle has been turned to **application position** causing the reducing-valve pipe pressure to flow to the application-cylinder pipe, and through it in one direction to the automatic brake-valve where that pipe is blanked under the rotary valve, and to the distributing valve where it actuates the application mechanism (nearing the latter, the pipe is given the color of the application-cylinder pressure); at the distributing valve the application-cylinder pressure fills the distributing-valve release pipe through which it only flows as far as the rotary of the independent brake-valve, being separated there, now, from its branch pipe connecting with the automatic brake-valve, the latter section of pipe being still open to the atmosphere through the rotary of the latter valve.

As the pressure of the reducing-valve pipe is being used as the actuating pressure, the check-valve in the fitting called "*strainer and check-valve*" prevents signal-pipe pressure from returning and so reducing as to cause the air-signal whistle to blow.

The action of the distributing valve resulting from the supply of reducing-valve pressure will be to cause main-reservoir air to flow to the brake cylinders of the locomotive, as indicated in color on the piping diagram. The reducing valve is commonly adjusted at 45 pounds, and whether the independent brake-valve is permitted to supply the full amount or only a portion of that pressure to the distributing valve, the latter will give an equal charge to the brake cylinders. Other than as stated, the pressure conditions remain as characterized throughout the piping in the usual **running position** of the brake-valves.

Referring to the distributing valve itself—Fig. 15 B

—it will be understood that the lower portion remains undisturbed, there being no reduction of brake-pipe pressure. The reducing-valve pressure, entering at II, passes directly to the application cylinder, *g*, and forcing piston 10 to the right, the operation of the application portion will be as usual in the cases heretofore described which should be generally understood by this time; the main-reservoir air from chamber *a* flowing to the locomotive brake cylinders through the brake-cylinder pipe connecting at CYLS. The application-cylinder pressure supplied from the reducing valve flows, *via* ports *h*, *k*, and *w*, to the application chamber which supplies the *volume* necessary in regulating a graduated application; through ports *h*, *k*, and *i*, it flows to the distributing-valve release pipe at IV; and through ports *h*, *s*, and *l*, the application-cylinder pressure also passes to the safety valve, which, in case the reducing valve should become defective and permit the full main-reservoir pressure to be supplied, would unseat itself and endeavor to reduce the excess of pressure over 68 pounds.

Independent Lap Position.

The application portion of the distributing valve will take the **lap position** following a locomotive-brake application by the independent brake-valve in any degree of pressure up to the full 45 pounds supplied by it from the reducing valve, for the reason that the braking air from the main reservoir is always at a higher pressure and will compel the lapping whenever the brake-cylinder pressure equals, or slightly exceeds, that in the application cylinder. The lapping movement of the application portion of the distributing valve is similar at all applications of the brake, and was explained in reference to Figs. 11 and 13; Fig. 16 shows it in the position of **independent lap.** It should be understood that this does not necessarily imply the lapping of the *independent brake-valve*—which latter would simply mean that the independent brake-valve had been so moved as to cut off the supply of reducing-valve pressure to the application cylinder.

In connection with this it should be noted that the equalizing portion of the distributing valve not having changed from its so-called *release and charging* position during the course of the independent application and resultant lap of the application valve, on that account the safety valve is still connected to the application cylinder in this position, and when it becomes necessary to hold the locomotive for some time with the independent brake-valve while standing, *that brake valve should not be returned to the* **lap position,** as a leak in the safety valve from scale or other obstruction between valve and seat would soon discharge the application-cylinder pressure and release the locomotive brake. Under the

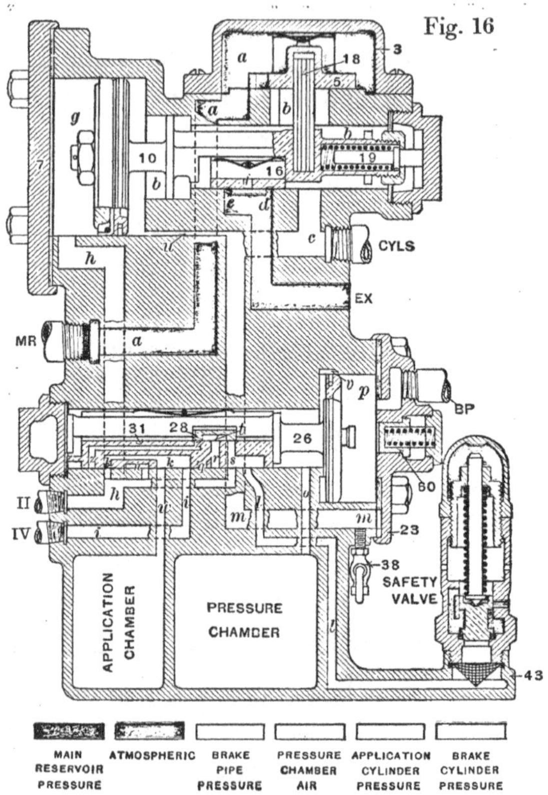

FIG. 16.—No. 6 Distributing Valve, Independent Lap Position.

circumstances always *leave the independent brake-valve in full* **application position,** and so long as it is there, and the pump is compressing air, you will have "a brake that won't come off."

Independent Locomotive-Brake Release, After Automatic Application.

When an automatic application of the brakes has been made, either service or emergency, by the automatic brake-valve, or from train-parting, or in any other manner whatever, the locomotive brakes can always be released, in full or in part, without interfering with the application of train brakes, by placing the independent brake-valve in **release position.** In the accompanying plate representing that phase of action, Figs. 17 A and 17 B, it is to be considered that an automatic service application of the locomotive and train brakes had been made by the automatic brake-valve which is now standing in the **lap position,** and the independent brake-valve is being held in its **release position.** It should be considered, in fact, that all parts of the locomotive-brake equipment had been in exactly the state as described in reference to Fig. 11, descriptive of **service-lap position,** and the locomotive brake-cylinder pressure was subsequently discharged by the independent brake-valve.

Referring to the piping diagram, Fig. 17 A, the independent brake-valve having been placed in **release position,** a port in its rotary valve has opened that terminal of the application-cylinder pipe to the atmosphere, thus discharging the application-cylinder pressure, and effecting the discharge at the distributing valve of the locomotive brake-cylinder pressure; although the piping for both pressures referred to is given the atmospheric color (*orange*), indicating that they have been entirely emptied, and the brake completely released, *independent release* may only mean a partial discharge of the application-cylinder pressure with a return to **lap position** of the independent brake-valve, and a partial release of the locomotive brake, as the latter can be graduated off—something impossible with the automatic brake.*

The distributing-valve release pipe is uncolored in the two figures of this plate, as it contains no pressure above atmospheric, being cut off from any pressure connections at the distributing valve, and is denied the color indicative of atmospheric air (orange), because it is also cut off from the latter by the rotary of the automatic brake-valve, and is intermediately

---

* An exception to this statement should be allowed, as the lately devised "Type L, High-Speed, Quick-Service," passenger triple-valve can be graduated off after an application, by partial recharges of the brake pipe; but this advantage is only to be taken when all of the cars in the train are equipped with the Type L triple valves.

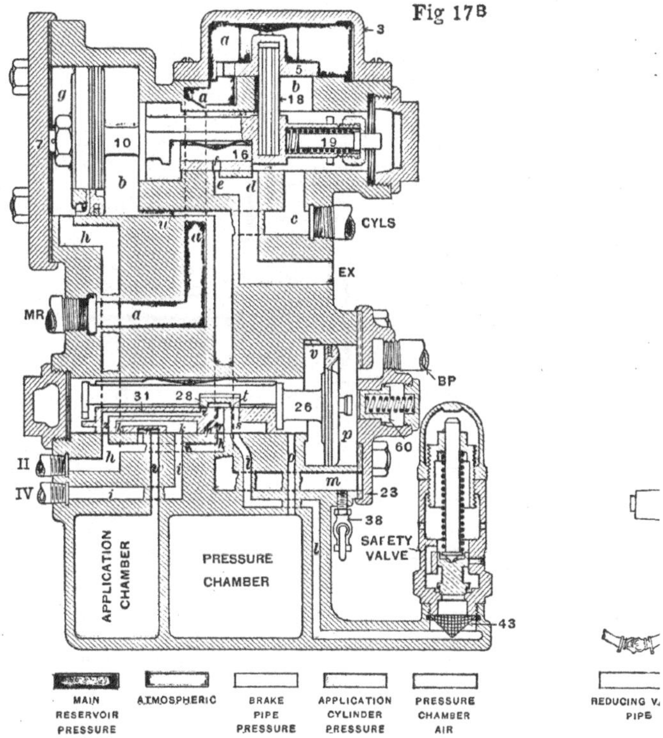

FIG. 17 B.—No. 6 Distributing Valve. Independent release position after automatic application.

Fig 17A

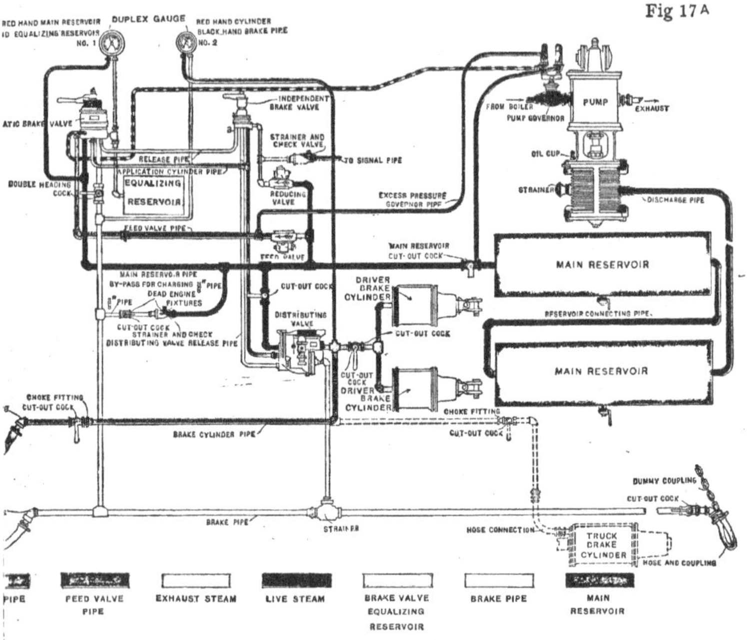

gram No. 6 E T Equipment. Colors showing sequence of pressures at independent locomotive-automatic application. Independent brake-valve in release position, automatic brake-valve in en colors show reduced, or less than normal, pressures.

ig 17B

pendent
ion.

Fig 17A

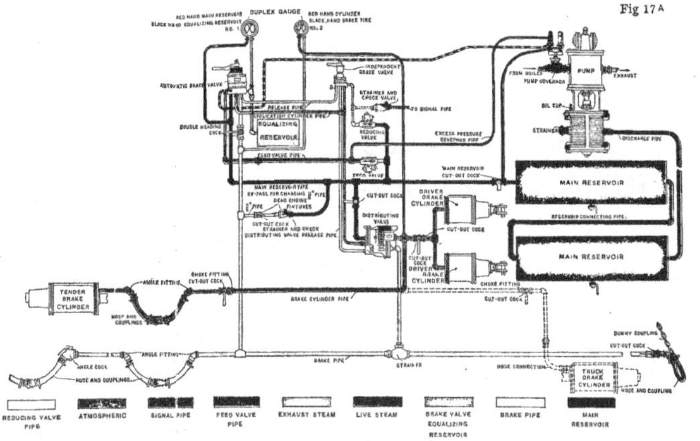

FIG. 17 A.—Piping Diagram No. 6 E T Equipment. Colors showing sequence of pressures at independent locomotive-brake release after automatic application. Independent brake-valve in release position, automatic brake-valve in lap position. Broken colors show reduced, or less than normal, pressures.

blanked by the rotary of the independent brake-valve.

Following an automatic service-application, if a graduated, or partial, release of the locomotive brake is made by the independent brake-valve the reduced braking pressure will remain constant (unless automatically increased by brake-pipe leakage), and if the release is complete the locomotive brake will stay off, under the conditions; but after an emergency application by the automatic brake-valve, any of the application-cylinder pressure that may be discharged by the independent brake-valve will be resupplied through the maintaining port in the automatic brake-valve, and the brake-cylinder pressure will be built up as fast as it is released, after the independent brake-valve has been lapped; so, to independently release the locomotive brake under *emergency* conditions, it is necessary to *hold* the independent brake-valve in **release position** as long as the automatic brake-valve remains in **emergency-application position.**

Fig. 17 B shows the distributing valve in the position of **Independent Release;** it will be unnecessary to allude to the lower portion any further than to note that it holds the same position as is usual at **service lap,** which was explained in connection with Fig. 11, having no effect on the independent release, nor would it make any particular difference if the equalizing portion had taken the emergency position, as shown in Fig. 12 B. The discharge of pressure from the application cylinder at the left of application piston 10 has resulted in the piston being moved to the left, as shown, by the pressure in chamber *b* on the right, thus opening the exhaust ports *e*, and *d*, through which the brake-cylinder pressure is exhausted to the atmosphere; if all of the application-cylinder pressure should be discharged, the application piston will remain in the **release position** as shown in the plate, and the brake-cylinder pressure will be fully released; but if the former pressure should be only partially discharged and the independent brake-valve returned to **lap position,** as soon as the brake-cylinder pressure in chamber *b* is exhausted to slightly less than the pressure remaining on the left of the piston, the application mechanism will be moved to the right until the exhaust ports, *e*, and *d*, are closed, and then will stop in the lap position, in which case all parts of the distributing valve will have assumed the same positions as they were shown in Fig. 11—**automatic service lap.**

THE QUICK-ACTION DISTRIBUTING VALVE.

NO. 6 DISTRIBUTING VALVE WITH QUICK-ACTION CYLINDER CAP.

With the advent of all-air-braked trains it became necessary to supply additional functions to the plain triple-valves referred to earlier in this work; no changes were found necessary in its action in response to service reductions, but at emergency applications on long, air-braked trains, the forward brakes would apply with full force before the brakes on the rear cars had started to set, resulting in the sudden "bunching" of trains with disastrous effects; in other words, the slack between the cars would close-in more rapidly than the brakes could serially apply throughout the train from front to rear.

Then the *quick-action triple valve* was evolved for freight and passenger cars, in the service action of which there was no difference from that of the plain triple-valve; but, following a quick, heavy reduction, the new triple was devised to discharge a portion of the brake-pipe air, and—further—to make this discharge into the brake cylinder, before the latter could receive any appreciable amount of auxiliary-reservoir pressure, resulting in an increased braking pressure at emergency as well as securing the full-on application of the rearward brakes before the slack could run in.

It has been found desirable in recent years to apply quick-action triple-valves to locomotive tenders, and this is particularly necessary when in case of double-heading, the engineer of the leading locomotive operates the brakes, as, if both engines and tenders are equipped with plain triple valves it is quite commonly impossible for him to secure quick action of the train brakes, owing to the large volume of brake-pipe air to be reduced between his brake-valve and the quick-action triple on the first car, and the resistance to air flow due to the several short bends and possible elbows in the brake pipe of both locomotives. As the E-T locomotive brake depends upon the distributing valve for the regulation of brake-cylinder pressure, and the graduating portion of the ordinary distributing valve acts on the same principle as the plain triple-valve, it follows that it is just as necessary that the distributing valve should possess the same quick-action feature of brake-pipe air-vent at emergency applications; on the theory, however, that some railroads do not encourage the double-heading of trains, the Westinghouse Air Brake Company furnish the distributing valve of the latest improved E-T equipment, No. 6 type, without the quick-action device, unless such shall be specified when ordering.

Fig. 18

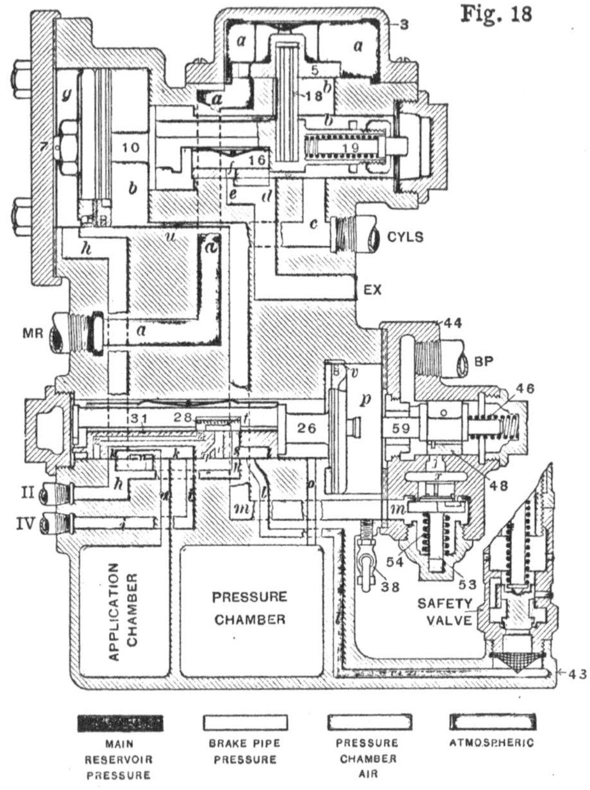

FIG. 18.—No. 6 Distributing Valve with Quick-Action Cylinder Cap. Released and charging position.

Quick-Action Cylinder Cap

Displacing the plain cap, or head, of the lower portion of the distributing valve and substituting the "Quick-Action Cylinder Cap" is all that is necessary to change the No. 6 equipment to a quick-action locomotive brake, and Figs. 18 and 19 show the distributing valve with the improved attachment.

NO. 6 DISTRIBUTING VALVE WITH QUICK-ACTION CYLINDER CAP.

RELEASE AND CHARGING POSITION.

Fig. 18 represents the distributing valve with QUICK-ACTION CYLINDER CAP, with all operative parts in the released and charging position; the conditions that exist within the distributing valve are the same as were described in connection with Fig. 9 B, and are, presumably, well understood; the piping diagram of Fig. 9 A will also apply as an adjunct to this distributing-valve chart, and need not be reproduced.

Referring to the quick-action cylinder cap, the graduating spring, 46, and the graduating stem, 59, appear, as in the plain cylinder-cap, except that stem 59 is lengthened and made to engage the emergency slide-valve, 48, between its shoulders; with the graduating spring relaxed within its limit, as shown, the slide valve covers port *j* in the seat, and brake-pipe air entering at BP only fills the slide-valve chamber of the cylinder cap, flowing thence into chamber *p* of the distributing-valve proper, and, as usual, through feed-groove *u* to the pressure chamber.

The provision of passageway *m* that carries the brake-cylinder air down to the lower part of the equalizing portion of the distributing valve is now seen to have

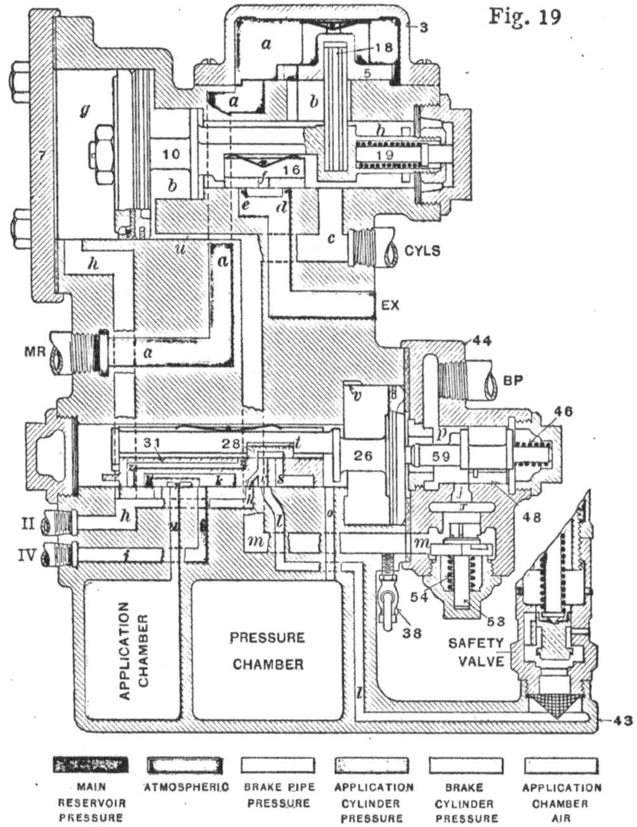

FIG. 19.—No. 6 Distributing Valve with Quick-Action Cylinder Cap. Emergency position.

been for the purpose of affording brake-cylinder pressure connection with the quick-action cylinder cap in the possible case of its use; passage *m* opens into the lower chamber under the rubber-seated, emergency check-valve 53, the check-valve being held to its seat in the absence of brake-cylinder pressure by the spring, 54. The intermediate chamber, *x*, is closed against either brake-pipe pressure or brake-cylinder air, now, and is therefore given no reference color.

At graduated service applications, equalizing piston 26 moves only to contact with graduating stem 59, without compressing the graduating spring or moving the emergency slide-valve, and the brake-cylinder pressure can only fill the check-valve chamber of the cylinder cap; so that no unusual results are obtained, except from emergency applications.

Emergency Position.

Fig. 19 represents the distributing valve with quick-action cylinder cap in **emergency position.** When a sudden, heavy reduction of brake-pipe pressure is made, the effect on the equalizing and application portions of the distributing valve is precisely the same as was described with reference to Fig. 12 B, **emergency position;** but the full stroke of the equalizing piston, compressing the graduating spring in the quick-action cylinder cap, carries emergency slide-valve 48 with it and uncovers port *j* in the slide-valve seat; brake-pipe air from chamber *p* rushing down through port *j* fills chamber *x* where its pressure unseats the emergency check-valve, 53, and flows to the passage, *m*, thence to passage *c* and the locomotive brake-cylinders through the pipe connecting at CYLS.

This described action takes place at the instant piston 26 strikes the quick-action cylinder-cap gasket, and, as main-reservoir air can not flow to the brake cylinders until—as the result of piston 26's stroke—pressure-chamber air has filled application cylinder *g* and forced piston 10 to the right to unseat application valve 5—it is obvious that brake-pipe pressure is the first to reach the brake cylinders; with the opening of the application valve the supplying pressures commingle in passage *c* as brake-cylinder air, which, when it becomes equal to the lowering pressure of the brake pipe, will permit check valve 53 to seat and prevent the brake-cylinder air from flowing back to the brake pipe through the open port, *j*; and, as soon as the brake-cylinder pressure becomes as great as the pressure in the application cylinder, piston 10 will close application valve 5, this upper portion assuming the lap position in the same manner that has been described repeatedly before.

Results from the quick-action cylinder cap produce no effect in the rest of the locomotive-brake equipment to differ from the action where the plain cylinder cap is used; merely, in the former case there is a more rapid reduction of brake-pipe pressure, and a consequent economy in the use of main-reservoir pressure. A piping diagram to accompany Fig. 19 would be exactly the same in outline and reference colors as Fig. 12 A.

In the case of quick-action triple valves, the vent of brake-pipe air at emergency applications will give a higher brake-cylinder pressure than can be obtained by a full, service application, and it may be imagined that because the No. 6 distributing valve gives a higher brake-cylinder pressure at an emergency than at a full, service application, that this, also, is due to the air received from the brake pipe; such is not the case, however, as the same difference in pressures is obtained with the use of the plain cylinder cap; and the reason for the increased pressure at emergency was explained in connection with Figs. 12 A and 12 B—before the quick-action attachment to the distributing valve was taken up. Regardless of from how many sources the brake-cylinder pressure is obtained, it can not become greater than the pressure in the application cylinder, and when it approaches an excess over that, the exhaust valve (16) will reduce it to an equalization with the application-cylinder pressure.

Service brake-pipe reductions of 25 pounds or more will cause the equalizing portion of the distributing valve, and the emergency slide-valve, to assume the same positions as represented in Fig. 19, but by the time the brake-pipe air is so far exhausted as that, the brake-cylinder pressure in passage *m* and the check-valve chamber will be equal to it, or greater, and check valve 53 can not be unseated. To secure full emergency action, the application must be made with brake cylinders (previously) empty; *partial emergency* is obtainable only so long as the brake-pipe pressure is appreciably greater than that in the brake cylinders; but after a service reduction of as much as 15 pounds, no quick-action results may be expected.

THE E-6 SAFETY VALVE.

Used on the No. 6 Distributing Valve.

Fig. 20 is an enlarged sectional view of the E-6 SAFETY VALVE that has been shown attached to the distributing valve in each of the preceding colored charts of the No. 6 E-T equipment. It is a quick-closing valve seating with a "pop" action, unlike the ordinary safety valves, is sensitive in operation and responds to slight differences in pressure.

The names of the parts of the safety valve are: 2, Body; 3, Cap Nut; 4, Valve; 5, Valve Stem; 6, Adjusting Spring; 7, Adjusting Nut. In each of the distributing-valve charts a STRAINER, 43, is seen just under the safety valve, and this piece should be included in the parts, although not appearing in the individual plate, Fig. 20; its use is essential, to prevent loose scale or other matter from being carried to the safety valve and lodging on the seat to make a leak that would result in the loss of brake-cylinder pressure.

The valve, 4, is cylindrical in form, fitting neatly in the surrounding bush which acts as a guide, and is held to its seat by the compression of spring 6 between the shoulder of the valve stem and adjusting nut 7. When the air pressure beneath valve 4 becomes greater than the resistance of the spring, the valve raises from

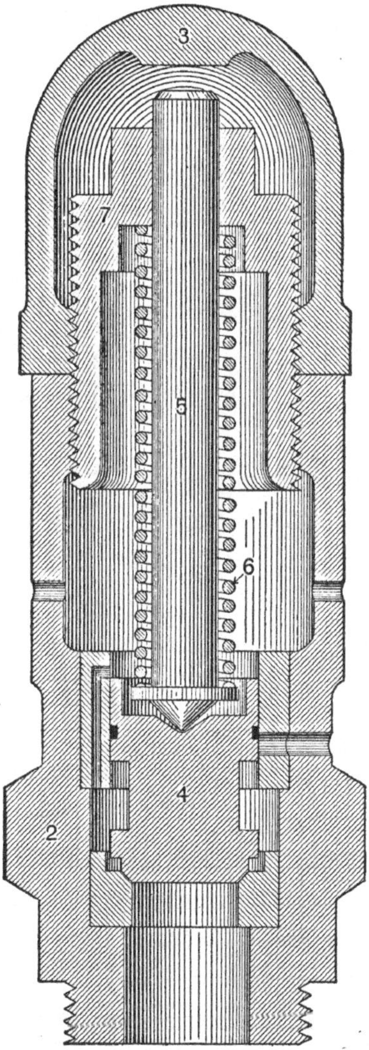

FIG. 20.—E-6 Safety Valve. Used on the No. 6 distributing valve.

its seat and then exposes a larger area to the air pressure which quickens its upward movement and prevents the valve from "chattering" on its seat.

Two ports are drilled in the valve bush upward to the spring chamber; and two are drilled outward through the bush and valve body to the atmosphere, although but one of each of these is shown in the cut. The lift of the valve is determined by the stem, 5, striking cap nut 3, when it closes the two vertical ports in the bush connecting the valve chamber and spring chamber, and opens the two lower ports to the atmosphere; as the exhausting pressure of the air below the valve becomes less than the pressure of the spring, the latter forces the stem and valve downward, during the movement of which the valve restricts the lower ports to the atmosphere and opens those between the valve and spring chambers, and the discharging air pressure then has access to the spring chamber; this chamber is always connected to the atmosphere by two small holes through the body, 2, and the air from the valve chamber, entering more rapidly than it can escape through these holes, causes pressure to accumulate above the valve and assist the spring to close it with the "pop" action before mentioned.

This safety valve used in connection with the No. 6 distributing valve should be adjusted for 68 pounds; this is done by removing cap nut 3 and screwing adjusting nut 7 down to raise, or up to lower, the pressure, and after the proper adjustment is made cap nut 3 must be replaced and securely tightened, and tested by operating with pressure a few times. The adjustment is more easily and accurately done on a shop testing rack.

The safety valve requires some attention and care. Particularly, it must be seen to that the holes in the valve body are always open, *but they must not be reamed out* by those who erroneously imagine that the capacity for pressure discharge should be increased—especially as to the two upper holes.

Occasionally the safety valve should be removed from the distributing valve, the strainer taken out and cleaned and the air passage that leads to the safety valve blown out by placing the independent brake-valve in application position. The safety valve should be taken to a bench and cleaned; remove the cap and adjusting nuts, the spring and stem; then invert the body and shake out the valve, 4, being careful that it shall fall on nothing hard that may dent or burr it; clean the inside of the body and the several holes referred to, and the valve bush; clean the valve, and rub a thin film of *graphite air-brake and triple-valve grease* around its sides and on the valve bush; replace everything, readjust to the correct pressure as before explained, re-attach the safety valve to the distributing valve, and test.

This should be done whenever the safety valve is taken off for adjustment, but the cleaning should not be delayed on that account. Situated where the distributing valve usually is, a great deal of gritty dust enters the safety valve through the small holes and, finding its way in between the periphery of the valve and the valve bush, becomes ground in and causes the valve to stick, or act irregularly.

The final test for pressure adjustment should be made with cap nut 3 screwed down tight against the safety-valve body as, with the cap removed, when the valve lifts from its seat it can rise so high as to close the lower ports from the valve chamber to the atmosphere, as well as the vertical ports through the bush; in which case the only discharge of pressure will be that which can leak around the sides of the valve.

ENGINEER'S BRAKE-VALVES OF THE NO. 6 E-T EQUIPMENT.

THE H-6 AUTOMATIC BRAKE-VALVE.

Fig. 21 is a photographic view of the H-6 AUTOMATIC BRAKE-VALVE with *pipe bracket*, complete, and Fig. 22 shows the same valve separated from its pipe bracket; for, like all of the other most important valves of the E-T equipment, the brake-valve proper can be removed for inspection or repair without disturbing any of the pipe joints. Fig. 23 shows two views of the brake-valve, with the addition of a plan, or transparent top view, of the rotary valve; the upper view of the brake-valve is taken from the top, on a section through the rotary-valve chamber, the rotary valve being removed; the lower one is a vertical section. In these views the pipe connections are indicated. Fig. 24 is a top view of the brake-valve, charting the different positions of the operating handle.

Referring to Fig. 24 and beginning at the left, we have **Release Position;** use of this position should only be made when the brake-pipe pressure has been reduced below the normal charge, and it is desired to release the train brakes; it has the effect of connecting the main reservoir directly with the brake pipe, and after it is believed that the brake-pipe pressure has been increased sufficiently to release all car brakes the handle must be moved to the second position, and it should never be left in **release position** long enough for the brake pipe to charge above the normal pressure—

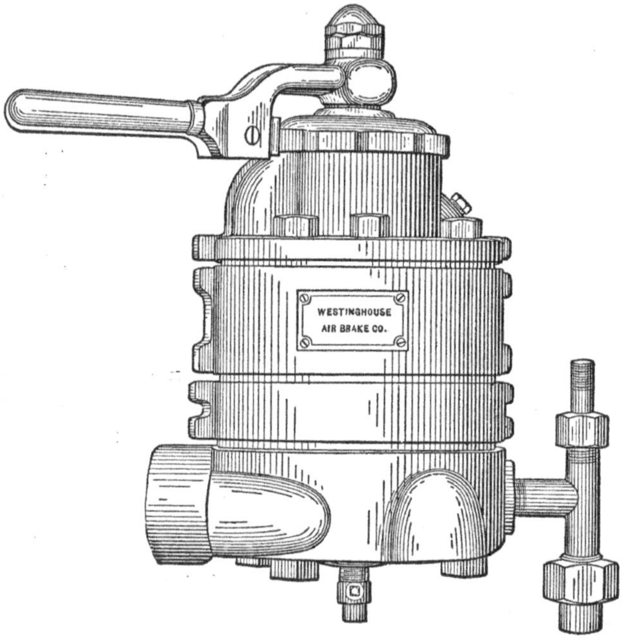

Fig. 21.—H-6 Automatic Brake-Valve. Complete.

usually 70 pounds. The locomotive brake, however, will not be released by the recharge of the brake pipe.

In the second, or **Running, Position** the locomotive brake will be released and held so, and this is its normal carrying position; the direct connection from main reservoir to brake pipe is now cut off, and air is supplied to the brake pipe from the 70-pound, feed-valve pipe.

In the third, or **Holding, Position** the 70-pound

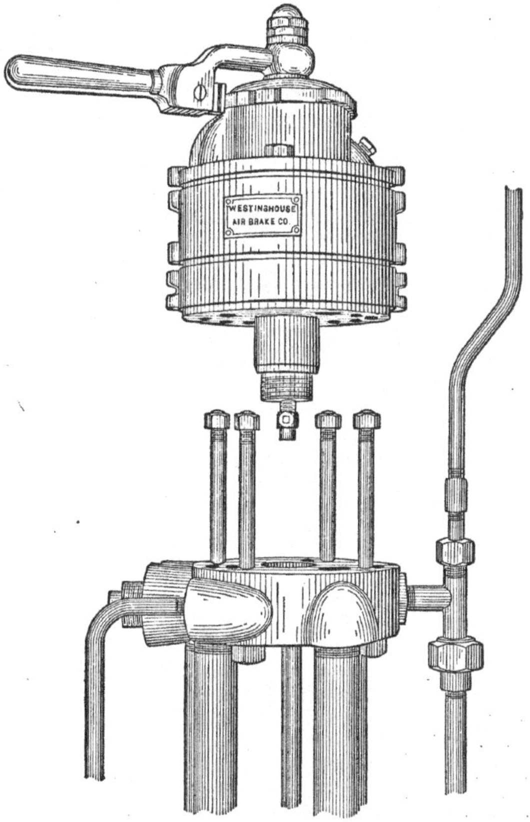

Fig. 22.—H-6 Automatic Brake-Valve. Removed from its pipe bracket.

pressure supply to the brake pipe is continued; it is, in effect, another *running position*, except that the release ports of distributing-valve and locomotive-brake cylinder pressures are closed. In returning the brake

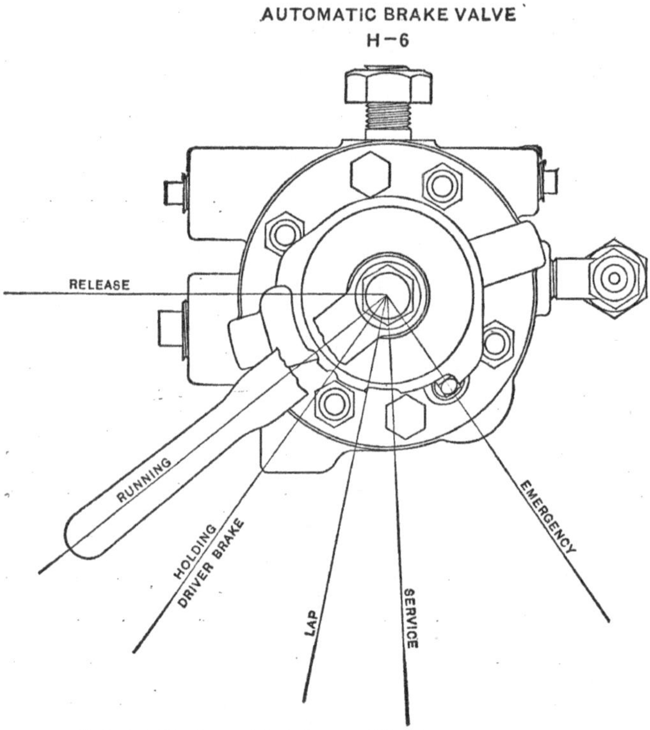

FIG. 24.—Positions of Automatic Brake-Valve Handle.

valve from **release position** it is a good plan always to go quickly to **holding position,** direct; this permits time for the afterthought as to whether it is best to release the locomotive brake at once, or to hold it on and so keep the head end from surging ahead until the rear brakes of the train have had ample time to fully release; the latter results will be attained by keeping the brake-valve in the **holding position** for a few moments, and meanwhile the brake pipe, the car auxiliary reservoirs, and the pressure chamber of the locomotive distributing-valve are receiving the normal pressure-recharge. When it is desired, the brake-valve handle can be moved back to **running position** and the locomotive brake will release.

In the fourth, or **Lap, Position** all separable connections in the brake-valve are blanked; this is a negative position, and is to be taken after a graduated, or service, reduction of brake-pipe pressure has been made, to hold the brake conditions in a fixed state; pressure is no longer being supplied to the brake pipe, and there is no further reduction of the latter, except from leakage.

Service Application is the fifth position; feed of air to the brake pipe is still cut off, and the brake-pipe pressure is being reduced through the *equalizing-discharge valve* of the brake-valve, but so gradually that the sensitive quick-action triple valves on the cars, and the distributing valve on the locomotive, will apply their respective brakes with a degree of force commensurate with the amount of reduction, yet not permit emergency action. When the desired reduction has been made—as indicated by the black hand of the large duplex gauge—the handle is to be returned to the **lap position.**

If the quickest and most powerful action of the brakes is desired, the brake-valve handle should be turned to **Emergency Application,** which is the sixth and last position to the right on the brake-valve; there is no supply of air to the brake pipe, and the pressure of the latter is discharged through a very large port in the brake-valve, causing such a quick and heavy reduction as to throw all triple valves and the distributing valve into emergency action; the brakes of a train of any length will apply at emergency much quicker than the slack can be run in solid from the rear end.

In the E-T equipment the brake-valve has been simplified in its functions over the brake-valve of the common, automatic system in which the *feed valve* was an integral part. In the H-6 brake-valve there are but two operative parts to study: the *rotary valve* that is operated by the brake-valve handle, the duty of which is to distribute the pressures that flow through the brake-valve; and the *equalizing-discharge valve* that automatically measures the discharge of brake-pipe pressure during a service application. An understanding of the operation of these two parts will give the air-brake student the knowledge that is necessary in regard to Westinghouse brake-valves of all types—if we add an explanation of the brake-pipe *feed-valve,* which will follow later.

Referring to Fig. 23, lower view: in this vertical, sectional cut, main-reservoir pressure is shown entering the lower piece of the brake-valve body termed the PIPE BRACKET, 5; the pipe connection referred to can be made from either the side or the bottom, the unused opening plugged; the supporting stud (No. 25, upper view) is in this lower piece, making it a permanent attachment to the boiler or other rigid base; and as the other three parts that form the body of the brake-valve are bolted together independently of the pipe bracket, the brake-valve proper may be removed without the disconnection of pipe joints, as the latter are all made to the pipe bracket direct.

This necessitates a number of ports through the sections of the brake-valve body, and the intermediate gaskets; thus the main-reservoir pressure is shown passing up through each and filling the TOP CASE, 4, in which the pressure covers the ROTARY VALVE, 6, In this plate the brake-valve is represented with the rotary valve in **running position,** in which the main-reservoir pressure goes no further than the top and sides of the rotary valve, except through a small port in the rotary valve and seat to the pipe connection, as shown in the upper view, leading to the *excess-pressure governor top.*

Fig. 23

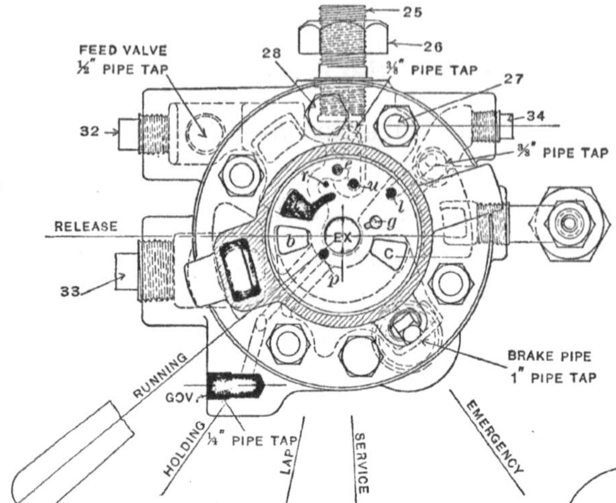

ROTARY VALVE SEAT.

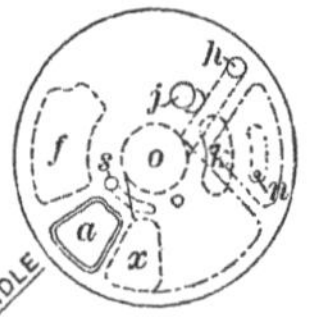

TRANSPARENT PLAN OF ROTARY VALVE.

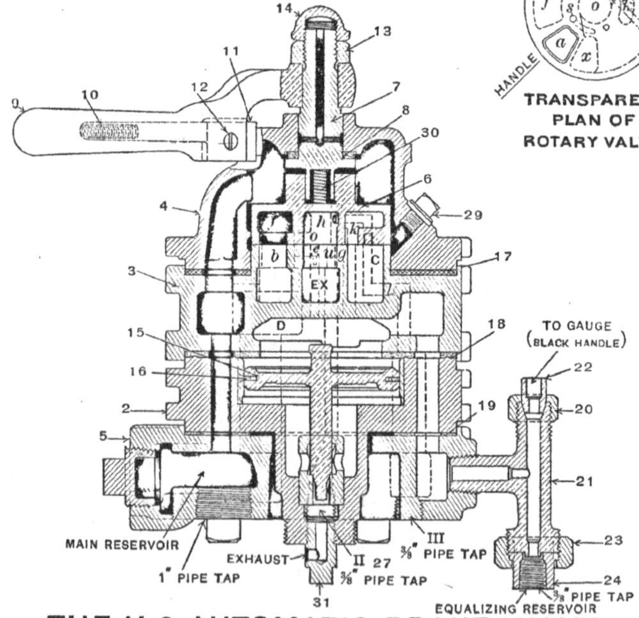

THE H-6 AUTOMATIC BRAKE VALVE.

RUNNING POSITION

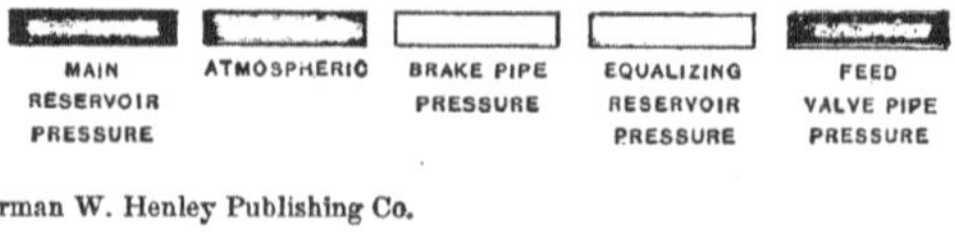

The connections through the automatic brake-valve, as established in the different positions of the rotary-valve handle, will be made plainer in the views subsequently to be given that will represent the rotary valve as transparent, and in the course of the description the reader will be referred back to this plate occasionally. The highest plane of the middle piece of the brake-valve body, 3, forms the seat of the rotary valve, and this casting is catalogued in whole as ROTARY-VALVE SEAT. The lower portion of the brake-valve proper —piece 2—is called BOTTOM CASE, and contains the EQUALIZING-DISCHARGE PISTON AND VALVE, 15. The three large gaskets making the joints between the three sections of the brake-valve proper, and the pipe bracket, are named as follows: 17, UPPER GASKET; 18, MIDDLE GASKET; and 19, LOWER GASKET. The large cavity in the center of the rotary-valve seat, marked EX, opens to the atmosphere through the large exhaust port leading out through the back side of the brake-valve body (this port is shown in dotted lines in the upper, plan, view).

To lubricate the rotary valve, remove OIL PLUG 29 before main-reservoir pressure is pumped up, and pour a little high-grade machine oil in the hole; it will fill the small recess around the inside of the top case at the converging edges of the rotary valve and seat. With this brake-valve, it is also possible to keep the leather KEY-WASHER, 8, soft and well lubricated, by removing handle lock-nut 14 and dropping some machine oil into the hole that is drilled down through the center of the ROTARY-VALVE KEY, 7; the oil fills the transverse port that is drilled clear through the handle key, and when main-reservoir pressure is off the rotary valve it seeps down between the washer and bearing surfaces of the rotary-valve key and top case, lubricating a point that has usually been neglected, and that when dry and gummy offers a greater resistance to the turning movement, sometimes, than a dirty rotary valve does.

Rotary-valve spring 30 holds the rotary valve and the key apart from each other, and to their seats, in the absence of main-reservoir pressure in the brake-valve, and this has the good effect of keeping dirt and scale from being blown on the seats when the pump is started. The HANDLE, 9, contains LATCH, 11, which fits into notches in the quadrant of the top case, so located as to indicate the different positions of the brake-valve handle; HANDLE-LATCH SPRING 10 forces the latch against the quadrant with sufficient pressure to indicate each position.

In referring to Fig. 22, it will be noticed that the removal of the four long bolts that go through the brake-valve as a whole will not permit the separation of the parts of the brake-valve proper. The plan view of the rotary-valve seat in Fig. 23 shows the location of these bolt-ends and nuts, 27, which must be taken off to remove the brake-valve proper from the pipe bracket; but to take apart the body sections of the brake-valve, for cleaning, oiling, etc., the cap screws, 28, must be removed, and these are shown as the two, plain, hexagon screw-heads, exactly opposite each other in the flange of the top case.

THE EQUALIZING-DISCHARGE VALVE.—If the engineman was to make all reductions of brake-pipe pressure directly through the rotary valve to the atmosphere, as he does in the emergency position, he would have to exercise an almost impossible skill to discharge the pressure rapidly enough with a long train, to get the pistons beyond the leakage grooves in the brake cylinders, and yet not fast enough to cause the quick-action triple valves to respond with their emergency action. The EQUALIZING-DISCHARGE VALVE, that has been an integral part of all Westinghouse brake-valves manufactured since 1890, automatically discharges the brake-pipe pressure during *service applications* at a rate of flow that is partly predetermined in the construction of the brake-valve, and partly governed by the volume of brake-pipe air that is being reduced; the number of brake-pipe pressure-pounds of the reduction is determined by the length of time the brake-valve handle is permitted to remain in the **service-application position**—with the older brake-valves, about 5 pounds per second.

In Fig. 23 the lower view of the brake-valve shows the equalizing-discharge mechanism very plainly; there is but one operating piece, the EQUALIZING PISTON, 15, with its PACKING RING, 16, the lower end of the piston stem forming the VALVE. Under the piston is brake-pipe pressure, and under the valve is atmospheric air in the "exhaust fitting." Above the piston is the air of CHAMBER D, which in the **running position** of the brake-valve is connected to brake-pipe pressure, and the pressures thus being equal on the top and bottom sides of the piston (at this time), it remains in the position shown with the valve seated.

(Although the *per-square-inch* pressures are equal on both surfaces of the piston, there is *more* pressure on the top than on the under side by just the area of the valve at the end of the stem that is exposed to the atmosphere; this slight difference in forces insures the proper seating of the valve.)

There must be substance to work on, always, and chamber D must have *volume*—something near a cubic foot of it—as well as pressure; but, to make it of that size the brake-valve would take up too much room in the cab; so, chamber D is made as small as possible—containing room merely for the necessary "lift" of the piston—and another chamber is provided elsewhere (usually outside the cab, under the running board), called the *equalizing reservoir*, in size 10-inch by 14½-

inch (formerly made 10-inch by 12-inch), which is in permanent connection with chamber D by a $\frac{3}{8}$-inch pipe leading from the lower union of the GAUGE AND EQUALIZING-RESERVOIR TEE, 21; and to the upper union of this tee, or fitting, 21, is connected the $\frac{1}{4}$-inch pipe to the large duplex gauge, the pressure of chamber D and equalizing reservoir registering by the black hand.

The legend "*black hand, equalizing reservoir,*" is on the face of this gauge in the No. 6 equipment; in all previous locomotive-brake equipments the legend read "*black hand, train line*" (brake pipe), because, in the **running position** of the brake-valve the equalizing-reservoir and brake-pipe pressures are always the same, and previous to the No. 6 equipment there was no other means of gauging the brake-pipe pressure.

Very few of the pipe connections and ports in the brake-valve are to be seen in Fig. 23. The feed-valve pipe connects to the pipe bracket as shown in the preceding piping diagrams, and its 70-pounds pressure (brown) comes up through port *d* in the rotary-valve seat (upper view, Fig. 23), flowing into cavity *f* in the face of the rotary valve, which, as is seen in the lower view, is also in register with brake-pipe port *b* in the seat; the brake-pipe connection to the pipe bracket is nearly under port *c* in the rotary-valve seat with which it is directly connected: note this in the upper view, and that a cavity through the interior of the rotary-valve seat (indicated by yellow, dotted semicircles) connects the facing ports, *b* and *c*, and through this route the feed-valve pipe pressure (brown) supplies the brake-pipe pressure (yellow). Cavity *k* in the face of the rotary valve now connects port *c* with another and smaller port in the seat—port *g*—that goes straight down into chamber D, and it is through this *equalizing port*, *g*, that the brake-pipe and equalizing-reservoir pressures are maintained at an equality while the brake-valve is in **running position.**

The movement of the brake-valve handle to **Lap Position** causes the rotary valve to blank all of its separable connections; continuing the movement to **Service-Application Position,** the blanked state of the ports referred to is maintained; there is no feed of pressure to the brake pipe. Cavity *h* is cored through the rotary valve from its annular opening that faces the valve seat to the central port, *o*, that is directly over the exhaust port, EX, in the center of the rotary-valve seat; in service-application position port *h* in the valve face is in register with the preliminary-exhaust port, *e*, in the rotary-valve seat that leads down into chamber D, and it is through this connection that the equalizing-reservoir pressure is discharged as the first step in the service discharge of brake-pipe pressure.

By noting closely, it will be seen in both views of the

rotary-valve seat that port *e* contains a bush just below the level of the seat, with a comparatively small opening drilled through it that so restricts the flow of air from chamber D that the pressure from the 10-inch x $14\frac{1}{2}$-inch equalizing reservoir will be reduced at the rate of about 5 pounds for each 2 seconds that the brake-valve handle remains in the service-application position.

As the pressure in chamber D begins to reduce, the brake-pipe pressure under the equalizing piston becomes the greater and the piston is forced upward, unseating the equalizing valve through which the brake-pipe air discharges to the atmosphere until its pressure is as low as, or slightly less than, the pressure remaining in chamber D, when the piston will be forced downward again, reseating the equalizing-discharge valve and terminating the reduction.

Although the time taken by the engineman in making a reduction of the equalizing-reservoir pressure always amounts to the same number of seconds to cause a given number of pounds reduction of brake-pipe pressure, regardless of the length of the train, after returning the brake-valve to the **lap position** the discharge of the latter pressure will continue for a time proportionate to the volume of brake-pipe air; understanding this, and by noting results, the engineman will be able to make a close guess as to the number of air-braked cars under his control by the duration of the brake-pipe exhaust.

The equalizing-discharge feature plays no part in making an emergency application, although in the **emergency position** of the brake-valve the equalizing-reservoir pressure is exhausted; in that position the brake-pipe air is discharged directly to the atmosphere through large ports in the rotary valve and seat, and this operation will be explained in reference to the rotary-valve charts to follow.

The arrangement of all ports in the rotary valve and seat, and air passages through the body of the brake-valve, should be thoroughly understood before proceeding further than Fig. 23, with the preceding piping diagrams for reference in remembering the pressure connections of the brake-valve.

The (brown) port, *d*, in the rotary-valve seat is directly connected with the feed-valve pipe, and the (yellow) port, *c*, with the brake pipe, and the flow of air from the former to the latter in the **running position** has been explained; the passages leading from ports *c* and *d* down to the pipe connections with the pipe bracket are shown in the dotted lines. The very small port, *r*, in the seat is drilled down to the atmospheric exhaust port, and when the brake-valve is in **release position** this *warning port*, *r*, is connected with port *d* by another port in the face of the rotary valve, and the sound of the escaping feed-valve pressure indicates to the engineman that the brake-valve is not in the normal carrying position. Port *u* in the seat connects with the application-cylinder pipe. Port *l* in the seat connects with the distributing-valve release pipe, and in **running position** is in register with port *h* in the rotary valve, through which the application cylinder of the distributing valve has temporary connection with the atmosphere *via* the large, central ports, *o* and EX, in the rotary valve and seat, respectively. The small (red) port, *p*, in the rotary-valve seat, quite close to the large, central port, EX, leads downward and out to the connection with the pipe to the diaphragm-valve chamber of the excess-pressure governor top; it is given the red color because in **release, running** and **holding positions** it receives main-reservoir pressure through port *s*, which is drilled through the rotary valve and has an extended cavity in the valve face through which the connection in the several positions is maintained.

Of the ports in the rotary valve not yet alluded to, *a* is cut clear through the valve, and in **release position** is located directly over port *b* in the seat, permitting main-reservoir air from over the rotary valve to flow into the large passages to the brake pipe by exactly the same route, from port *b* onward, as was taken by the feed-valve pressure in **running position.** Port *x* is a large opening in the face of the rotary valve, with a wide, shallow cavity cored out of the interior of the valve —fan-shaped, as indicated by the dotted lines—that reaches inward to the central exhaust port, *o;* when the brake-valve is in **emergency-application position** the facing port, *x*, is in register with the direct brake-pipe port, *c*, in the seat, this connection providing the heavy discharge of brake-pipe pressure to the atmosphere that insures quick action of the brakes. Port *j* goes through the rotary valve, and in **release position** registers with port *g* in the seat so that main-reservoir air will be temporarily supplied to chamber D and provide a holding-down pressure *above* the equalizing piston as great as the pressure flowing to the brake pipe *beneath* it.

As stated, port *k* is a cavity in the face of the rotary valve, but it has a thin, fan-shaped extension cored in the interior of the valve, shown by dot-and-dash lines, that connects with the small port, *n*, and the latter is widened and lengthened to form a larger cavity in the face of the rotary valve; when the brake-valve is in **emergency position** the facing cavity of port *n* connects with port *u* in the seat, the end of cavity *k* has connected with the narrow extension of the feed-valve port, *d*, in the seat, and port *j* is lying over port *d;* through this combination of ports main-reservoir pressure, entering port *j* at the top of the rotary valve, finds a connected passage to the application cylinder of the distributing valve as the *maintaining pressure* heretofore alluded to.

These last-mentioned ports have been somewhat changed in configuration in the rotary valves of more recent manufacture, but simply as an improvement in detail, and the maintaining-pressure supply remains unchanged. In the following representations of the rotary valve in its several positions the newer arrangement of those ports will be used, as the later design is easier to understand.

THE ROTARY VALVE.—The six plates that follow are plan views of the rotary-valve seat of the H-6 brake-valve, seen through a transparent rotary valve in its six operating positions. The **gray** tint represents the ROTARY VALVE, and the ports that are cut vertically through it from top to face show the ROTARY-VALVE SEAT as plain **white.** All ports and passages in the ROTARY VALVE are indicated in **red** outline; those that are cut clear *through* are in continuous red lines; **red,** dotted lines show ports and cavities in the *face* of the rotary valve; and **red** dot-and-dash lines indicate ports and cavities in the *interior* of the rotary valve. Ports and cavities in the ROTARY-VALVE SEAT are in **black** outline, the dot-and-dash lines indicating cavities below the surface of the seat.

Reference numbers and letters are avoided as far as possible in these transparencies; the names of the several ports in the rotary-valve seat are printed plainly thereon, but no words, figures nor letters appear in connection with the ports in the rotary valve, to prevent confusion, and further the appearance of its transparency; it is to be presumed that the student has already become familiar with the appearance and positions of these ports, and their names or reference letters; however, in the following explanations of the several positions of the rotary valve, when its ports may be alluded to by reference letters that have not been memorized, the student is referred back to Fig. 23. The rotary-valve handle is of the same **gray** color that represents the rotary valve, with **red** outline, and its operative positions are indicated by the six radial lines with the words **release, running, holding, lap, service,** and **emergency.**

In Fig. 25, as the handle shows, the rotary valve is in **Release Position.** The two large ports in the rotary-valve seat worded "to brake pipe" are connected by a large cavity beneath the surface, indicated by the broken, black lines in half circles, with the brake-pipe connection directly beneath the right-hand port; the large port that opens vertically through the rotary valve is now in exact register with the left-hand brake-pipe port in the seat, and as main-reservoir pressure is always present on the top of the rotary valve, its route in flowing directly to the brake pipe is plainly traced. The equalizing reservoir is also receiving main-reservoir pressure, as port *j* that goes through the rotary valve is directly over the port to chamber D in the seat; port *j* is extended as a short cavity in the face of the rotary valve, in order that main-reservoir pressure

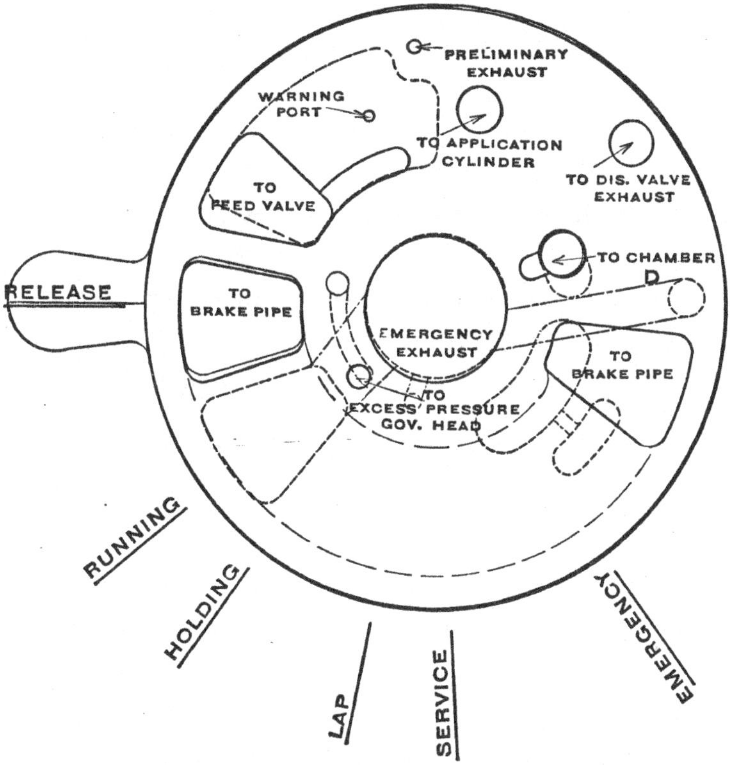

FIG. 25.—H-6 Brake-Valve. Top view of transparent rotary valve, and plan view of rotary-valve seat. Release position.

will continue to feed to chamber D when the brake-valve handle is moved toward **running position,** as long as the same pressure is flowing to the brake pipe—these are the *equalizing ports.* Port *s* is the third and smallest of the three that are cut through the rotary valve vertically; it lies to the left of and very close to the large, central exhaust cavity, with a connecting groove in the face of the rotary valve, the farther end of which in this position overlies the port in the seat that leads to the pipe connection with the excess-pressure head of the pump governor, and supplies the main-reservoir pressure that controls the action of the pump while the brake-valve handle is in **release position.** The very large cavity, *f*, in the face of the rotary valve now covers the feed-valve port in the rotary-valve seat, and overlaps a greater area of the seat that includes the *warning port* through which the feed-valve air discharges into the atmospheric outlet of the large, emergency exhaust port, giving warning of the possibility of brake-pipe overcharge.

Except for this connection with the warning port, in release position of the brake-valve the feed-valve pipe is blanked against any delivery of its air; and in case the feed valve leaked through—never so slightly—the feed-valve pipe would accumulate a pressure in excess of 70 pounds—possibly main-reservoir pressure—which would prevent the excess-pressure top of the governor from closing-off steam from the pump; this would be particularly undesirable in descending heavy grades, when it is sometimes desired to charge the train line to 90 pounds by leaving the brake-valve in release position. Any leak through the feed valve that would not be manifest in running position, however, will be disposed of by the blow-off through the warning port.

In Fig. 26 the brake valve has been moved to **Running Position,** the large supply-port, *a*, and the equalizing port, *j*, in the rotary valve are now blanked on the rotary-valve seat, cutting off the direct flow of main-reservoir pressure to brake pipe and chamber D. The large, adjoining ports in the seat—"*to feed valve*," and "*to brake pipe*," have been brought into conjunction by the very large cavity, *f*, in the face of the rotary valve, through which the brake pipe is now being supplied with air from the feed valve that limits it to 70-pounds pressure. The state of equalization of the brake-pipe and chamber-D pressures is still maintained: in this position, through cavity *k* in the face of the rotary valve, which connects the right-hand brake-pipe port in the seat with port *g*, also in the seat, that is worded—"*to chamber D*." In this, the latest model of the H-6 rotary-valve, port *k* is in permanent connection with a twin, but somewhat smaller, port, *n*, by a narrow groove in the face of the rotary valve, providing the same service in a much simpler and easier understood manner than the former arrangement; port *n*, however, is only brought into play in the position of **emergency application.** The long, tunnel-like cavity through the rotary valve from port *h* in the face to the large, central

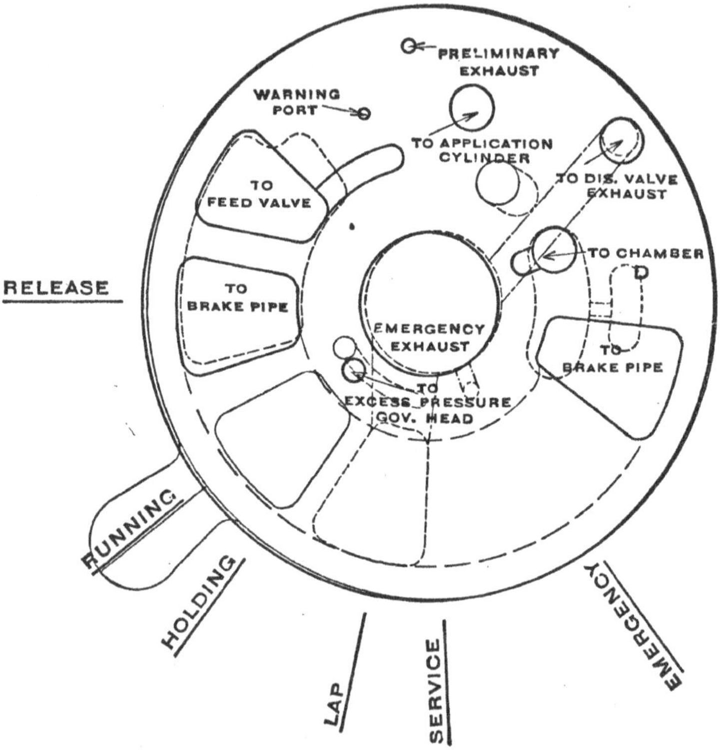

FIG. 26.—H-6 Brake-Valve. Top view of transparent rotary valve, and plan view of rotary-valve seat. Running position.

exhaust-cavity, now connects the latter with the port in the seat, "*to dis.-valve exhaust*"— that is, to the distributing-valve release pipe; and it is through this connection that the application-cylinder pressure is exhausted when the brake-pipe recharge forces the lower, or *triple-valve*, portion of the distributing valve into release position, and thereby effects the release of the locomotive brake. The port in the rotary-valve seat connecting with the pipe to "*excess-pressure governor head*" is still receiving main-reservoir pressure through port *s* in the rotary valve and its lengthened cavity in the valve face, and the low-pressure feature of the governor is still controlling the action of the pump. All other ports are blanked between the face and seat of the rotary valve.

The movement of the brake-valve handle from **running position** to **Holding Position** (Fig. 27), while shifting all ports in the rotary valve a short distance, does not separate any of the connections that were held in the former position—except one: the "*dis.-valve exhaust*" port in the rotary-valve seat is cut off from the atmospheric port in the center of the brake-valve by the shift of the formerly connecting port, *h*, which is now blind on the rotary-valve face. **Holding position,** thus becomes another **running position,** so far as the common functions of the brake-valve are concerned (although a lesser area of the feed-valve port is now opened in connection with the left-hand brake-pipe port in the seat), but it permits those functions while retaining the application of the locomotive brake, by the closing of the distributing-valve release port; to secure this holding effect, the brake-valve handle must be brought

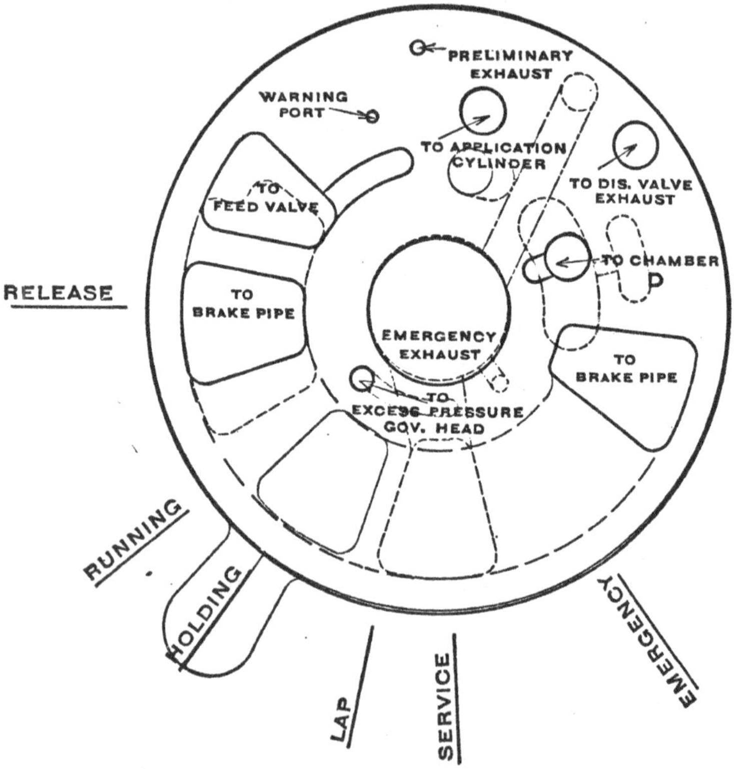

Fig. 27.—H-6 Brake-Valve. Top view of transparent rotary valve, and plan view of rotary-valve seat. Holding position.

from **release position** *directly* to **Holding Position,** passing the **running position** quickly so that none of the application-cylinder pressure will be discharged. The small port, *s*, through the rotary valve is now in exact register with the port in the seat connecting with the "*excess-pressure governor head*"; and we have found that in the three positions of the brake-valve in which pressure is supplied to the brake pipe, main-reservoir pressure flows to the excess-pressure top of the pump governor, with the effect of regulating that pressure at its minimum figure.

The position of the rotary-valve handle in Fig. 28 shows that the brake valve is in the **Lap Position.** No pressure is supplied to the brake pipe, as the large, main-reservoir supply port through the rotary valve is blind on the seat, and the large cavity, *f*, in the face of the rotary valve no longer connects the feed-valve and brake-pipe ports in the seat. Cavity *k* in the face of the rotary valve, while still lying over the port to chamber D, no longer connects the latter with the right-hand brake-pipe port in the rotary-valve seat. The small port, *s*, through the rotary valve has finally parted with the port in the seat that is designated "*to excess-pressure governor head*," thus closing off the supply of main-reservoir pressure to—and, in effect, cutting out—that portion of the governor, and permitting the pump to increase the main-reservoir pressure to the amount permitted by the high-pressure top of the governor. The location of port *j*, now, over the end of the channel that is an extension of the feed-valve port in the seat, is without results further than that it fills the feed-valve port, and pipe connecting,

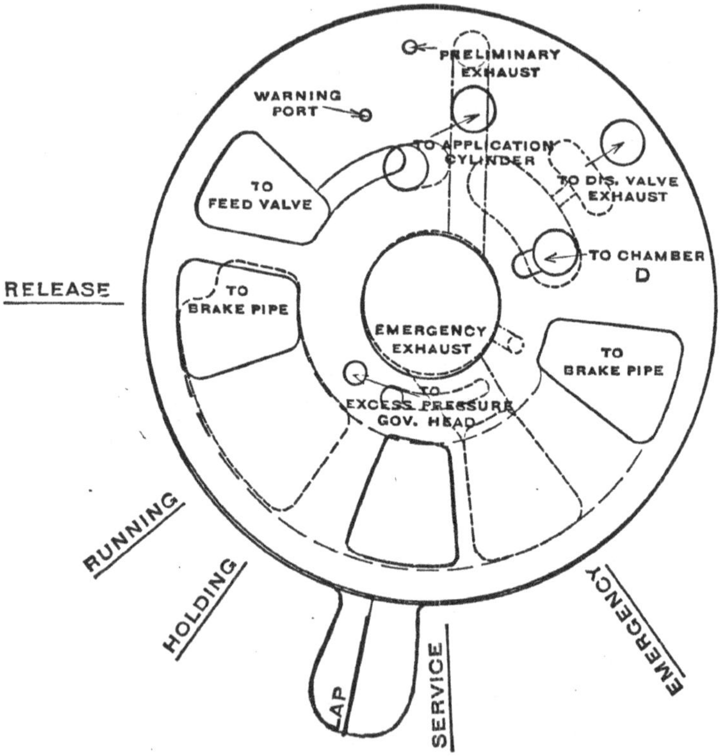

Fig. 28.—H-6 Brake-Valve. Top view of transparent rotary valve and plan view of rotary-valve seat. Lap position.

with main-reservoir pressure, and this introduces a possibility that was not contemplated: In the No. 6

equipment, port *p* in the rotary-valve seat and the pipe that connects it with the excess-pressure governor top could have been omitted, and the other small pipe from main reservoir direct to the high-pressure governor top should then have branched to the diaphragm-valve chambers of both governor tops; with the brake-valve in **service** and **emergency positions,** the delivery of main-reservoir pressure to the feed-valve port as just described in the **lap position,** continues, and in charging the regulating-spring chamber of the excess-pressure governor top with the maximum pressure that portion of the governor is cut out just as effectively as by the blanking of port *p* in the rotary-valve seat.

The short movement of the brake-valve handle from the position of **lap** to **Service-Application Position,** represented in Fig. 29, does not change the relations of the ports in the rotary valve and seat that existed in the former position, except that it brings port *h* in the face of the rotary valve directly over the "*preliminary exhaust*" port in the seat; equalizing-reservoir pressure from chamber D now flows upward through the preliminary-exhaust port into port *h*, and through its long, connecting cavity in the interior of the rotary valve to the large, central exhaust cavity, and thence to the atmosphere; as heretofore explained, this reduction of equalizing-reservoir pressure causes an equal pressure reduction of the brake-pipe air, and when the former has been sufficiently reduced—as indicated by the black hand of the large duplex gauge—the brake-valve handle must be returned to the **lap position,** to

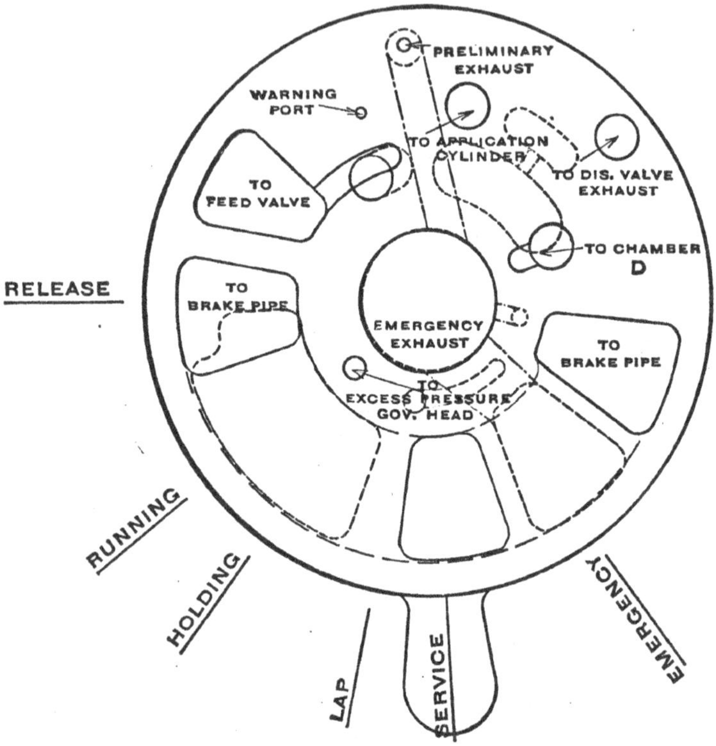

FIG. 29.—H-6 Brake-Valve. Top view of transparent rotary valve, and plan view of rotary-valve seat. Service-application position.

stop the further discharge of equalizing-reservoir pressure and limit the power of the application.

Service-application position is usually taken, for the initial reduction, by bringing the brake-valve handle directly there from the **running position**; but after re-

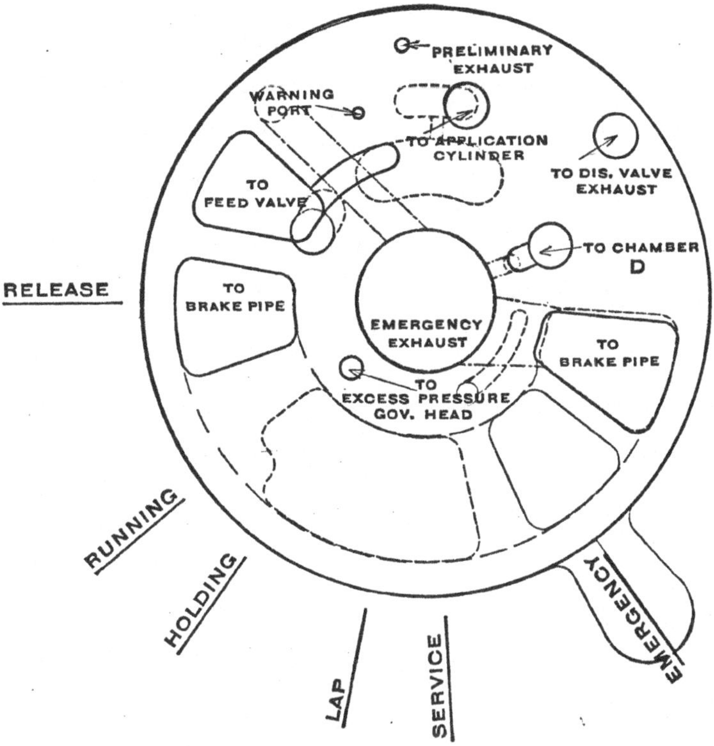

FIG. 30.—H-6 Brake-Valve. Top view of transparent rotary valve, and plan view of rotary-valve seat. Emergency-application position.

turning the handle to **lap position,** any further reductions will start from that position, as described.

Rotary Valve. Emergency Application

The sixth and last position of the brake-valve handle toward the right, is, as shown in Fig. 30, **Emergency-Application Position,** in which the effect is to apply the brakes with heaviest force and quickest action, and is usually taken direct from **running position.** Same as in the last two preceding positions, all supply of pressure to the brake pipe is closed off; the excess-pressure governor top is still cut out from main-reservoir connection; port and passage *h* to the atmospheric exhaust cavity in the center of the rotary valve is no longer in connection with the preliminary-exhaust port in the seat, but the equalizing-reservoir pressure is being discharged through the small port in the face, and its short, interior cavity of the rotary valve that connects the port in the seat "*to chamber D*" with the large, central "*emergency exhaust port*" leading out to the atmosphere. The principal feature in this position is the heavy discharge of brake-pipe pressure directly to the atmosphere through the rotary valve; the large cavity, *x*, in the face of the rotary valve now lies in even register with the right-hand brake-pipe port in the rotary-valve seat—the port closest to the brake-pipe connection; this port, *x*, is continued as an interior cavity cored within the rotary valve that connects with the central "*emergency exhaust*" port, and hence the name of the latter, for the brake-pipe pressure takes this route through the rotary valve to the atmosphere at emergency applications.

The *maintaining pressure* that is supplied to the application cylinder of the distributing valve, when the brake valve is in **emergency position,** has a tortuous passage in getting through the rotary valve; the main-reservoir pressure upon the rotary valve passes through it *via* port *j*, and gives the maximum pressure charge to the feed-valve port in the seat, a condition that was initiated in the **lap position,** as explained in connection therewith; but now the rotary valve has turned far enough that port *k* in the face of the rotary valve is overlapping the end of the channel branching from the feed-valve port in the seat and receiving *main-reservoir pressure* therefrom, which flows from port *k* to its twin port, *n*, through the small, restricting-port that connects these two concentric ports in the face of the rotary valve; from port *n* the maintaining pressure flows into the port in the seat designated "*to application cylinder*," which connects with the application-cylinder pipe leading to the distributing valve.

The small port connecting the twin grooves, *k* and *n*, in the face of the rotary valve is known as the *blow-down timing port*, as in addition to building up the application-cylinder pressure at emergency applications—raising it several pounds above the 65 pounds at which the pressure chamber and application cylinder equalize

—it serves at emergency applications of the High-Speed brake to so restrict the blow-down of application-cylinder pressure as to give approximately the same time between the maximum and minimum that is obtained in the use of the high-speed reducing valve.

THE S-6 INDEPENDENT BRAKE-VALVE.

Fig. 31 is a top view of the S-6 INDEPENDENT BRAKE VALVE used in the No. 6 E-T LOCOMOTIVE-BRAKE EQUIPMENT, with a diagram of the five operating positions of the rotary-valve handle; Fig. 32 is a photographic view of the brake-valve complete; in Fig. 33 the valve is shown removed from its pipe bracket; and Fig. 34 represents a sectional view of the complete brake-valve, with another view from the top as a plan of the rotary-valve seat, and a transparent plan view of the rotary valve.

Like the automatic brake-valve, the two body sections of the independent brake-valve are held together by cap screws, independent of the longer bolts that bind the brake-valve proper to the pipe-bracket section; and in general construction the two brake-valves are quite similar, including the handles, facilities for oiling, etc.

Referring to Fig. 31, the farthest position of the handle to the left is **Release,** and it is only necessary to use this position when the automatic brake-valve is in some other than **running position,** and it is desired to release the locomotive brake; the **release position** should be used, however, *whenever the locomotive brake remains applied undesirably* and the automatic brake-valve *is in* **running position,** as may happen on the second engine in double-heading. **Release position** discharges the application-cylinder pressure directly to the atmosphere through the exhaust port in the center of the bottom of the independent brake-valve; the handle will not remain in **release position** unless held there, a coil spring within the housing of the valve body

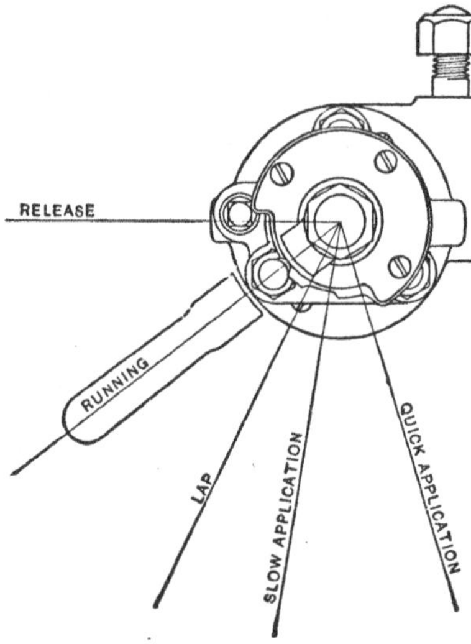

FIG. 31.—S-6 Independent Brake-Valve. Positions of rotary-valve handle.

turning the rotary valve to **running position** when the hand is removed from the handle.

Running Position is second, and the position in which the handle is standing as shown in the cut; it should always be left in this position, except when the independent brake is applied, as a port in its rotary valve connects the distributing-valve release pipe through to the automatic brake-valve, thus *admitting* of the release of the locomotive brake in **running position,** *if the automatic brake-valve is also in* **running position.** Ordinarily, therefore in operating the independent locomotive brake its release is secured simply by replacing the independent brake-valve in **running position.**

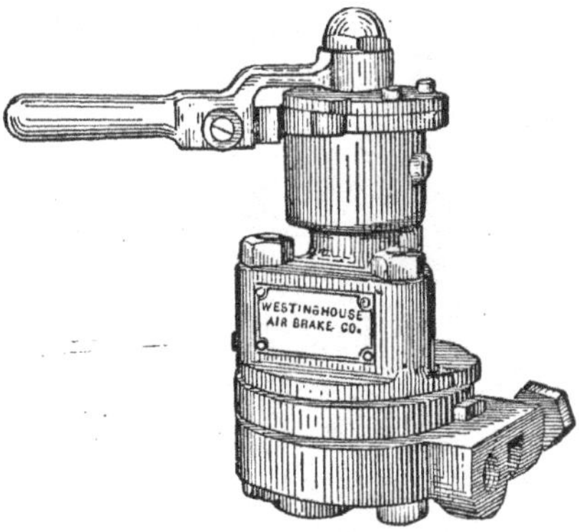

FIG. 32—S-6 Independent Brake-Valve, Complete.

In the third, or **Lap, Position,** the distributing-valve release pipe routing is discontinued beyond the rotary of the independent brake-valve, all the ports of which are blanked except the receiving port for reducing-valve pressure; in this condition, at an automatic application nothing unusual would be effected; but the locomotive brake could not be released by returning the automatic brake-valve handle to **running position,** until the independent brake-valve should be replaced in **running position,** also.

The fourth is **Slow-Application Position;** a very small port in the rotary valve permits reducing-valve pressure to flow quite gradually to the application cylinder of the distributing valve, permitting very fine graduating of the power applied to the locomotive brake.

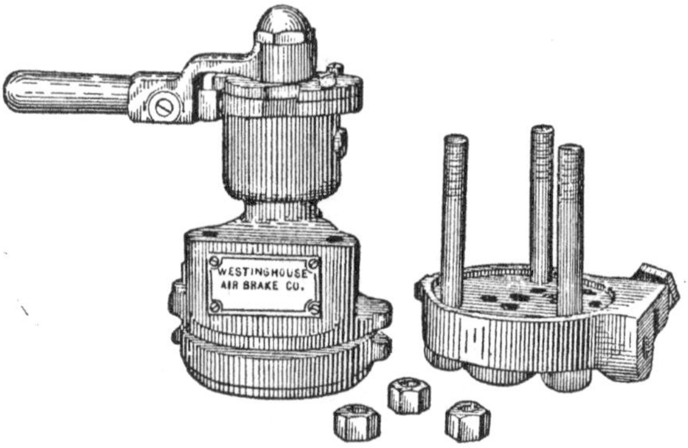

FIG. 33.—S-6 Independent Brake-Valve. Removed from pipe bracket.

In the fifth, or **Quick-Application Position,** the supply of reducing-valve pressure to the distributing valve is given quickly, through a larger port in the rotary valve, and as it results in a sudden and heavy application of the locomotive brake this position should be used with judgment; there is but little danger of the rotary valve being moved to this position when the **slow-application position** is intended, as, when the handle passes

Fig. 34

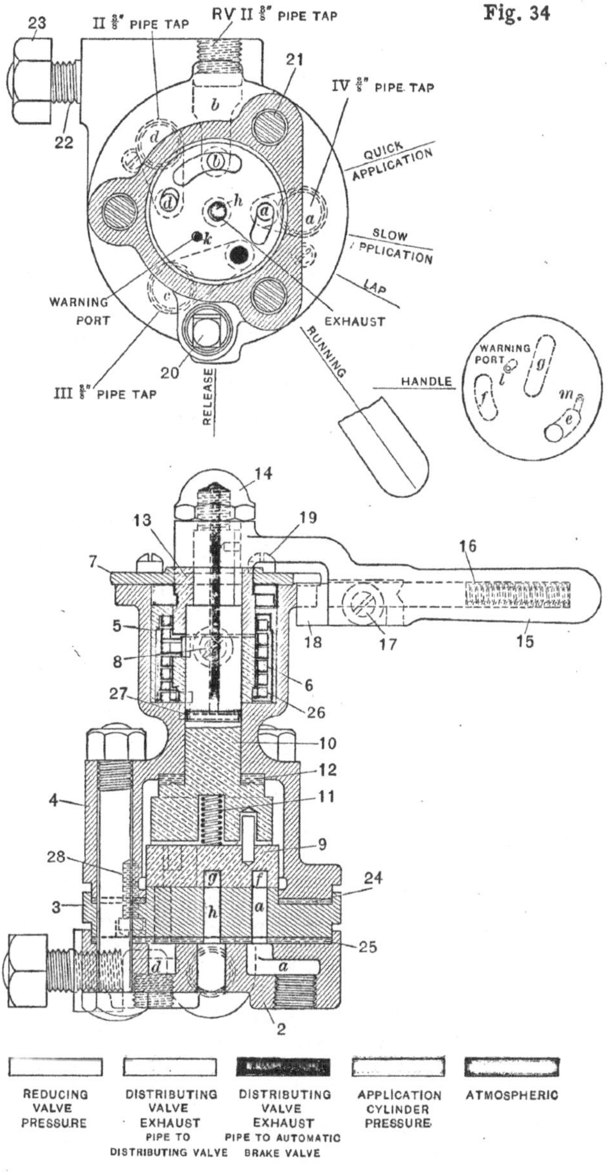

Fig. 34.—S-6 Independent Brake-Valve. Sectional elevation, plan view of rotary-valve seat, and transparent plan view of rotary valve.

the latter position the coil spring within the valve body is again encountered; some force is necessary to bring the handle to the **quick application position,** and when the hand is removed the brake-valve will rotate back to the **slow-application position.**

Like the automatic brake-valve, the S-6 Independent Brake-Valve is removable for repair, etc., without any pipe joints having to be disconnected. Fig. 32 shows the valve complete; and in Fig. 33 it is shown separated into the brake-valve proper, and the lower section, or pipe bracket, which carries the supporting stud-bolt.

Fig. 34 shows a vertical section through the INDEPENDENT BRAKE-VALVE, and a horizontal section through the valve body on the plane of the rotary-valve seat, with rotary valve removed and shown aside as a *transparent*, top or plan view. In the sectional views the pipe connections and positions of the handle are indicated. The names of the parts are:

2, PIPE BRACKET; 3, ROTARY-VALVE SEAT; 4, VALVE BODY; 5, RETURN-SPRING CASING; 6, RETURN SPRING; 7, COVER; 8, CASING SCREW; 9, ROTARY VALVE; 10, ROTARY-VALVE KEY; 11, ROTARY-VALVE SPRING; 12, KEY WASHER; 13, UPPER CLUTCH; 14, HANDLE NUT; 15, HANDLE; 16, LATCH SPRING; 17, LATCH SCREW; 18, LATCH; 19, COVER SCREW; 20, OIL PLUG; 21, BOLT AND NUT; 22, BRACKET STUD; 23, BRACKET-STUD NUT; 24, UPPER GASKET; 25,

LOWER GASKET; 26, LOWER CLUTCH; 27, RETURN-SPRING STOP; 28, CAP SCREW.

Unlike the automatic brake-valve, the air-pressure supply does not pass up through a port in the different sections of the INDEPENDENT BRAKE-VALVE BODY and come upon the rotary valve direct; this *reducing-valve pressure* (45 pounds), from its pipe connection with the bracket section, flows up through a passage to port *b* and its channelled extension in the rotary-valve seat that connects with port *e* in the rotary valve in all positions; port *e* includes a groove in the face of the rotary valve and a port extending vertically through it, by means of which the *reducing-valve pressure* flows to the top of the rotary valve at all times, thus holding it to its seat. Port *a* leads to that section of the *distributing-valve release pipe* that goes to the distributing valve (connection IV—distributing-valve charts), and port *c* leads to the other section of this pipe that goes to the automatic brake-valve (connection III—Fig. 23). Port *d* leads to the *application-cylinder pipe* to the distributing valve (connection II—dist.-valve charts). Port *h* in the center of the rotary-valve seat is the exhaust port, leading directly down to the atmosphere beneath the brake-valve. The "*warning port,*" *k*, also leads to the atmosphere. The long, radial groove, *g*, in the face of the rotary valve is always in communication with the atmosphere through its permanent connection with the central exhaust port, *h*. Port *m* in the face of the rotary valve is connected with the channel, *e*, by a small, interior port. *F* is a channelled cavity in the face of the valve; and *l* is a port through the rotary valve from top to face, where it is extended as a short groove.

In Figs. 35, 36, 37, 38, and 39, the functions of the INDEPENDENT BRAKE-VALVE in its five operative positions are exemplified by views of the rotary valve as transparencies, and, through it, the rotary-valve seat. In connection with this study, reference should be made to the *piping diagram*, and *distributing-valve chart*, that represent the effects of operation of the brake-valve in each position as taken up.

The **red** lines indicate ports and cavities in the ROTARY VALVE, the unbroken lines representing ports that pass clear through the valve from top to face; dotted lines indicate cavities that are channelled out in the face of the valve; and dot-and-dash lines indicate passages in the interior of the rotary valve that are used to connect facing ports. The ports and cavities in the ROTARY-VALVE SEAT are shown in **black** lines.

Referring to Fig. 35, **Release Position** of the Independent Brake-Valve: the groove in the rotary-valve seat that is a part of the "*reducing-valve pressure*" port is always in connection with the large port, *e*, through the rotary valve, either directly or through the groove in the face of the valve, and its extended cavity as in Fig. 35, by means of which the reducing-valve pressure flows on top of the rotary valve, serving at present only to hold it to its seat, but ready for use when an independ-

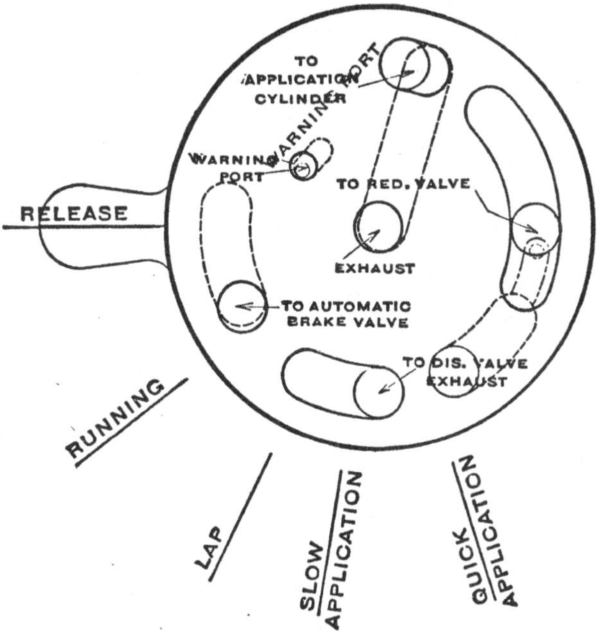

FIG. 35.—S-6 Independent Brake-Valve. Top view of transparent rotary valve, and plan view of rotary-valve seat. Release position.

ent application is made. The long, radial groove, *g*, in the face of the rotary valve is now connecting the "*application-cylinder*" port in the seat with the atmospheric exhaust-port *h* in the center of the valve seat, thus discharging the pressure from the application cylinder of the distributing valve *at the independent brake-valve*, and releasing the locomotive brake under any circumstances of its application, regardless of the position of the automatic brake-valve handle. The small "*warning port*" through the rotary valve is now in register with the similarly designated port in the seat, providing a slight discharge of reducing-valve pressure to the atmosphere, the noise from which serves to warn the engineer that the independent brake-valve is in a position in which the locomotive brake would immediately release after any manner of automatic application; the warning feature acting as a safeguard in case the *return spring* should become broken.

It will be noticed that the route of the *distributing-valve release pipe* from distributing valve to automatic brake-valve is broken in this position, by the separation of the two ports in the seat indicated as the independent brake-valve terminals of the piping—the sections leading to *distributing valve*, and *automatic brake-valve*, respectively.

In Fig. 36 the *return spring* has rotated the INDEPENDENT BRAKE-VALVE to **Running Position**; in this, the second operative position, ports *e* in the rotary valve, and *b* in the seat, are still in connection to provide the supply of reducing-valve pressure above the rotary valve; but the "*application-cylinder*" port and "*warning port*" in the seat are now blanked under the face of the rotary valve.

No positive work is performed by the independent

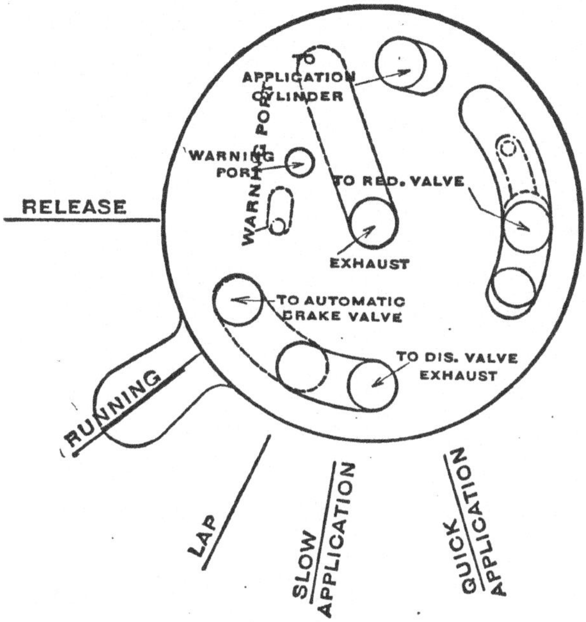

FIG. 36.—S-6 Independent Brake-Valve. Top view of transparent rotary valve, and plan view of rotary-valve seat. Running position.

brake-valve in this position; in ordinary switching service it is used to release the locomotive brake, but this is not positive, and depends upon the automatic brake-valve also being in **running position.** It is the regular carrying position of the handle, featured as the only position in which the *distributing-valve release pipe* has an open route through the independent brake-valve to the automatic brake-valve, the channel, *f*, in the face of the rotary valve of the former now connecting the two ports in the seat leading, as indicated, to the two sections of the pipe running to the distributing valve, and automatic brake-valve, respectively. As it is only through this port connection that locomotive-brake release can occur when the automatic brake-valve is placed in **running position**; it is very important that the independent brake-valve shall be carried in **running position,** *always*, when not being used.

It is apparent that the INDEPENDENT BRAKE-VALVE can release the locomotive brake in *one position* under any circumstances, and in *two positions* when the automatic brake-valve is in **running position**; in the latter case, after an independent application the return of the independent brake-valve to **running position** permits the application-cylinder pressure—which at that time is in the *distributing-valve release pipe* as far as the INDEPENDENT BRAKE-VALVE—to pass through the continuation of that pipe to the automatic brake-valve, and through its rotary valve to the atmosphere; brake-cylinder pressure being equally discharged at the exhaust port of the distributing valve.

The third position of the INDEPENDENT BRAKE-

VALVE is **Lap,** as represented in Fig. 37. This is a negative position, in which all ports are closed—except, of course, the ports for the reception and passage of reducing-valve pressure to the top of the rotary valve.

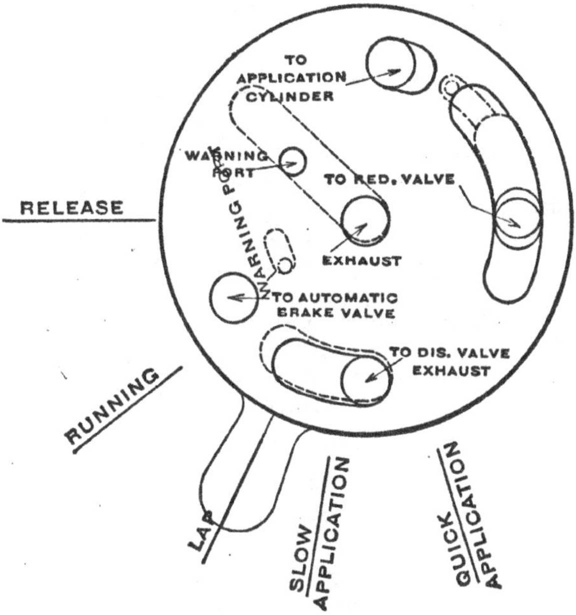

FIG. 37.—S-6 Independent Brake-Valve. Top view of transparent rotary valve, and plan view of rotary-valve seat. Lap position.

The *distributing-valve release pipe* is again cut off and blanked at the rotary valve, by the switching of the groove, *f*, in the face of the rotary away from the port "*to automatic brake-valve*" in the seat.

The object of **lap position** is to hold the locomotive braking conditions as they are after making a graduated independent application, and should be used at no other time than after graduating on, or graduating off,

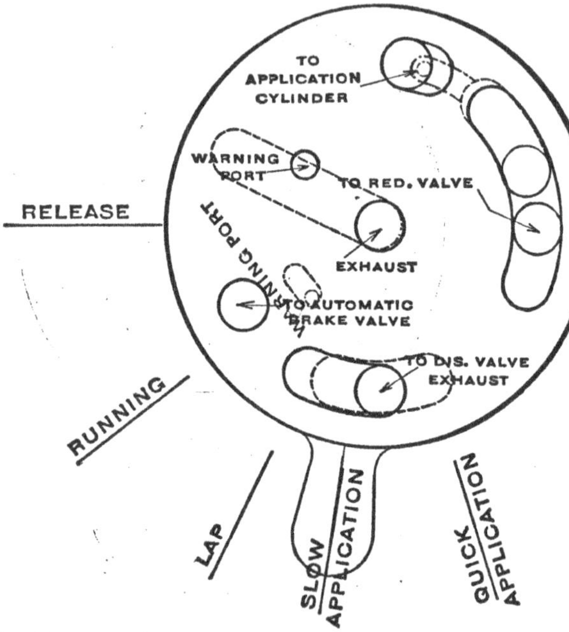

FIG. 38.—S-6 Independent Brake-Valve. Top view of transparent rotary valve, and plan view of rotary-valve seat. Slow-application position.

the locomotive brake when the automatic brake-valve is in **running position.**

Slow Application, the fourth position, is represented in Fig. 38. Port *e* through the rotary valve is still permitting reducing-valve pressure to flow to the top of the valve. The small port, *m*, in the face of the rotary valve is now in register with the port in the seat "*to application cylinder*"; through an interior cavity in

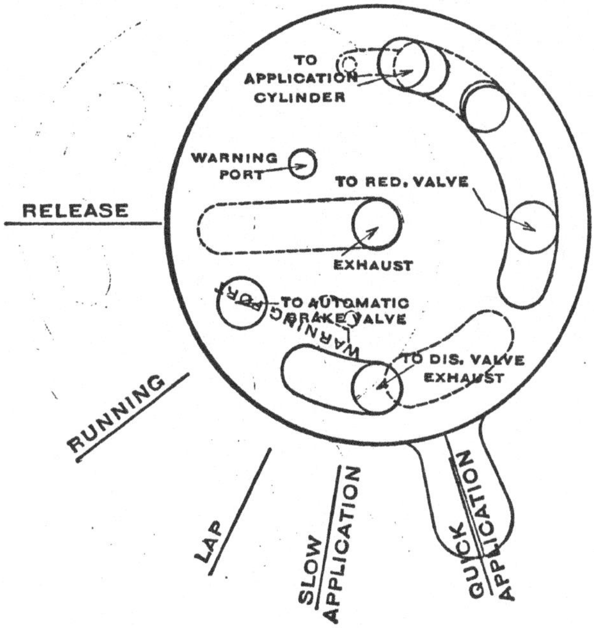

FIG. 39.—S-6 Independent Brake-Valve. Top view of transparent rotary valve, and plan view of rotary-valve seat. Quick-application position.

the rotary valve port *m* is connected with the groove of port *e*, permitting reducing-valve pressure to flow to the application cylinder of the distributing valve, in which it acts as previously described to apply the locomotive brake. The slowness of an independent application in this position is due to the very small size of port *m* through which the application-cylinder pressure is fed. All other ports in the rotary valve and seat remain as in the **lap position.**

The fifth and final position of the S-6 INDEPENDENT BRAKE-VALVE, as shown in Fig. 39, is that of **Quick Application.** The conditions are exactly the same as in the previous position, except that the rotary valve has turned far enough to bring the large groove of port *e* in its face into connection with the "*application-cylinder*" port in the seat, providing an enlarged passage for the flow of reducing-valve pressure to the application cylinder of the distributing valve, and effecting a quick application of the locomotive brake.

THE B-6 FEED VALVE.

The B-6 Feed Valve furnished with the No. 6 Equipment, photographic views of which are shown in Figs. 40 and 41, is the common *slide-valve feed valve,* the duty of which is to regulate pressure supply to the brake pipe, but improved by the hand-wheel regulating device and an enlarged regulating valve. In the ordinary automatic equipment, the feed valve was attached directly to, and was considered a part of, the automatic brake-valve; in all E-T equipment, it is located in the line of one of the two pipes that supply main-reservoir air to the brake-valve. The pipe that is directly supplied by the feed valve leads to the automatic brake-valve, and is called the *feed-valve pipe,* and in **Running** and **Holding positions** of the brake-valve it is in open port connection with the brake pipe.

All forms of Feed Valve are interchangeable. As originally designed for attachment to the G-6 automatic brake-valve, the feed valve hangs downward in its proper position; in its application to the previous styles of the E-T brake it was turned upside-down—sticking upward; while in the No. 6 equipment we find it again turned down in its rightful position; and the reasons follow.

There are two air ports side by side in the connecting face of the feed valve, and as we stand in front of the

G-6 brake valve the left one is the entering port for main-reservoir pressure, and the right one is the port of exit, or brake-pipe connection. As used in the E-T equipment, the feed valve is attached directly to a "*pipe bracket*" (note appearance in Figs. 40 and 41), and as

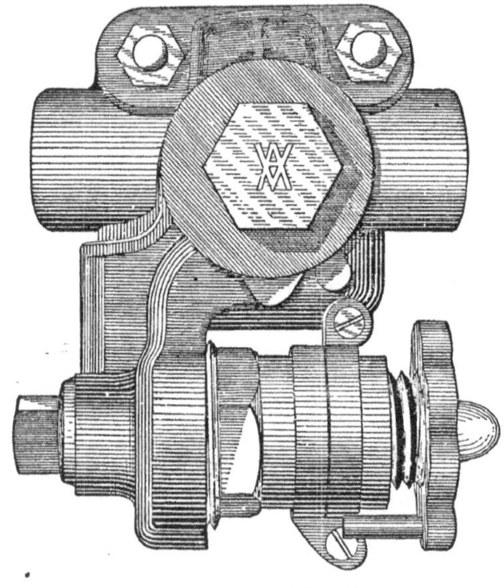

FIG. 40.—The B-6 Feed Valve. Valve and pipe bracket complete.

usually placed, the main-reservoir air enters the pipe bracket from the right. The pipe brackets of the No. 5 equipment were simply made—right-hand pipe connection leading to right-hand face port, and left-hand pipe connection to left-hand face port; this would have reversed the order of passing the air through the feed valve, but by *turning the valve upside-down,* the port connections were made to coincide, and they were so inverted in the No. 5 and all preceding E-T equipments.

The *crossed passage,* "F" pipe bracket is used in the No. 6 E-T equipment, and, as the name indicates, the passages in this pipe bracket *are crossed,* so that

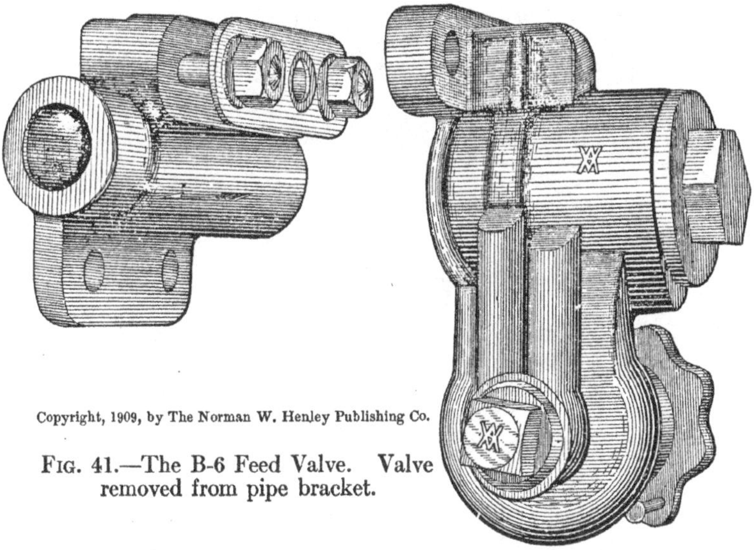

FIG. 41.—The B-6 Feed Valve. Valve removed from pipe bracket.

main-reservoir air entering at the right-hand pipe connection will pass to the left-hand port from which it enters the feed valve, leaving through the right-hand port to the left-hand pipe connection, which is to the feed-valve pipe leading to the automatic

Fig. 42

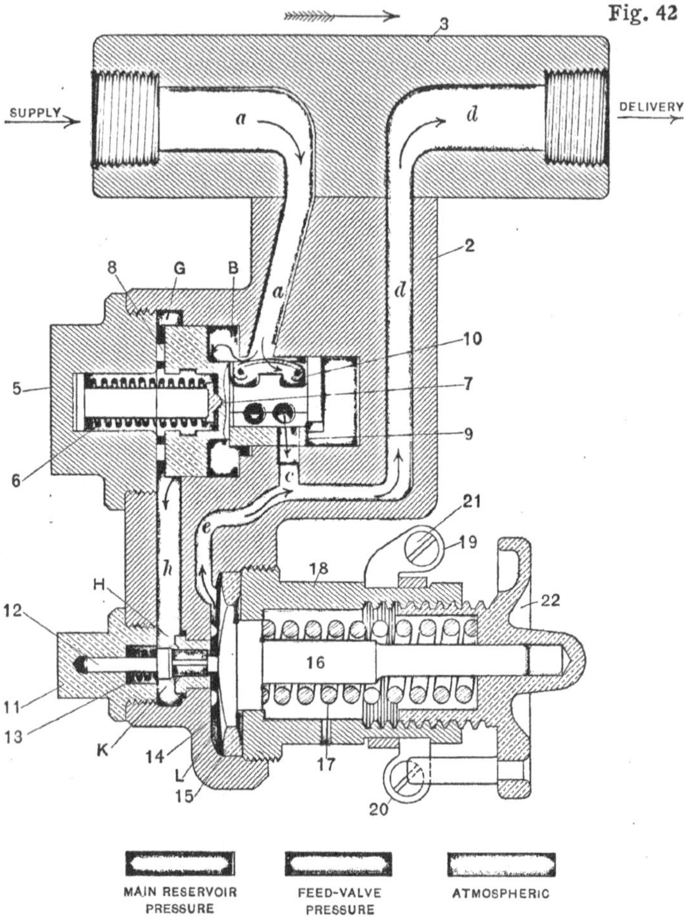

FIG. 42.—Diagram of B-6 Feed Valve. Open position. MR—Main-reservoir pipe. FVP—Feed-valve pipe.

brake-valve. In order that the feed valve may be correctly placed, in the No. 6 equipment, the new, crossed-passage pipe bracket has a lug cast upon it that interferes with the attachment of the feed valve unless it is placed right-side up.

The improvements in the B-6 FEED VALVE permit charging to the regulated pressure somewhat quicker, and maintaining the pressure more accurately than the old style did, under the variable conditions of short and long trains, and of good and poor maintenance. Also, the regulation can be quickly and accurately changed from 70 pounds to 110 pounds brake-pipe pressure, or the reverse, by turning the hand wheel until the pin strikes the opposite stop; or any other pressures as the minimum and maximum can be used, by adjusting the stops to secure the desired amount.

Figs. 42 and 43 are diagrammatic views of the FEED VALVE and PIPE BRACKET, having the ports and operating parts in one plane to facilitate description; and to simplify description the *direct passage* pipe-bracket is represented. The names of the parts shown in the diagrams are as follows: 2, VALVE BODY; 3, PIPE BRACKET; 5, CAP NUT; 6, PISTON SPRING; 7, PISTON-SPRING TIP; 8, SUPPLY-VALVE PISTON; 9, SUPPLY VALVE; 10, SUPPLY-VALVE SPRING; 11, REGULATING-VALVE CAP; 12, REGULATING VALVE; 13, REGULATING-VALVE SPRING; 14, DIAPHRAGM; 15, DIAPHRAGM

RING; 16, DIAPHRAGM SPINDLE; 17, REGULATING SPRING; 18, SPRING BOX; 19, UPPER STOP; 20, LOWER STOP; 21, STOP SCREW; 22, ADJUSTING HANDLE.

The feed valve consists of two sets of operating parts, the supply (upper), and regulating (lower). The supply parts, which control the flow of air through the valve, consist of the supply valve 9 and its spring 10, the supply-valve piston 8 and its spring 6. The regulating parts consist of the regulating valve 12, regulating-valve spring 13, diaphragm 14, diaphragm spindle 16, regulating spring 17, and regulating handle 22.

Referring to Fig. 42 in which the feed valve is represented in the OPEN POSITION, main-reservoir air enters through port *a*,*a* to the supply-valve chamber B, forces supply-valve piston 8 to the left, compresses piston spring 6, and causes the port in supply valve 9 to register with port *c* in the seat; this permits air to pass through ports *c* and *d* to the feed-valve pipe at FVP, the pressure of which flows through port *e* to diaphragm chamber L.

It will be observed that no packing is used in piston 8, as a certain amount of air leakage past it is desirable, and necessary to the proper operation of the valve; air feeding by the piston to the left can not accumulate in chamber G above feed-valve pipe pressure at this time, for regulating valve 12 is open and connects chamber G to the feed-valve pipe, through passage *h*, port K,

Fig. 43

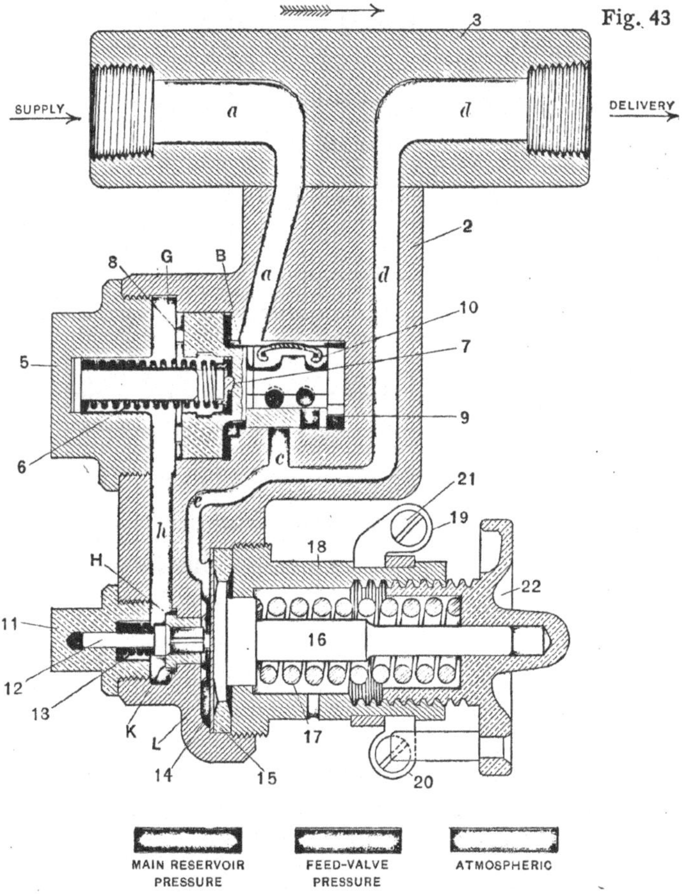

Fig. 43.—Diagram of B-6 Feed Valve. Closed position. Connections: See Fig. 42.

chamber L, and passage *e*, *d*, *d*. Regulating valve 12 is held open by the force of regulating spring 17 against the center of the diaphragm 14, upon which the regulating valve bottoms.

When the air in the feed-valve pipe reaches the pressure at which the feed valve is adjusted—say 70 pounds—its power upon the diaphragm in chamber L overcomes the pressure of regulating spring 17, which is then further compressed (refer now to Fig. 43, CLOSED POSITION), and this permits the small spring 13 to drive the regulating valve 12 to the right until it seats, closing port K and thus cutting off communication between chamber G and the feed-valve pipe; the feed of main-reservoir air from chamber B by piston 8 continues, and the pressure in chamber G quickly becomes equal to that of chamber B; the air pressures on both sides of piston 8 now balancing each other, piston spring 6 forces the piston and supply valve 9 to the right, closes port *c* and stops the flow of air to the feed-valve pipe.

When the feed-valve-pipe pressure begins to lessen, the regulating spring, 17, again the stronger, unseats valve 12, and the pressure of chamber G equalizes with that of the feed-valve pipe; spring 6 is of very light tension, and the drop of pressure in chamber G results in piston 8 again being forced to the left by main-reservoir pressure, supply valve 9 is opened and the flow of pressure to the feed-valve pipe is resumed; and so on.

In this type of feed valve the duplex adjusting arrangement eliminates the necessity of the two feed valves formerly provided for high- and low-pressure service. The spring box 18 has two rings encircling it, which are split through the lugs marked 19 and 20 in the diagram, and which may be secured in any position by the screw 21. The pin forming part of adjusting handle 22 limits the movement of the handle to the distance between stops 19 and 20. When adjusting the valve, stop 19 is located so that the compression of regulating spring 17 will give the desired high brake-pipe pressure, and stop 20 so that the spring compression is enough less to give the low brake-pipe pressure. After which, by simply turning handle 22 until its pin strikes either one of these stops, the regulation of the feed valve is changed from one brake-pipe pressure to the other.

To change to other minimum and maximum pressure-adjustment positions of the stops, slacken screws 21, which allows stops 19 and 20 to turn around spring box 18. Then turn the adjusting wheel handle 22 (compressing the regulating spring to increase the pressure; slackening the spring tension to decrease it); adjust for the lower pressure first, and turn until the valve closes at the minimum brake-pipe pressure desired, when stop 20 should be moved to contact with the handle pin, and securely fastened in that position by tightening its set-screw. Then turn wheel handle 22 clockwise until the higher adjustment is obtained, bring stop 19 in contact with the handle pin, and tighten set-screw 21. The stops are generally placed to give 110 pounds high, and 70 pounds low, brake-pipe pressures, the former being the pressure commonly accepted for High-Speed Braking, and the latter is the long-established pressure for all other conditions of air-brake service.

When replacing the feed-valve on its bracket after removal, the gasket, shown in Fig. 41, must always be in place between the valve and bracket, to insure a tight joint.

The FEED-VALVE PIPE: As has been stated, besides carrying the feed-valve pressure to the automatic brake-valve, this pipe has but one other connection—the branch pipe leading to the regulating-spring chamber of the *excess-pressure top* of the pump governor. When the automatic brake-valve is in **release position** the brake pipe does not receive feed-valve pressure, and to guard against the possibility of a slight leak through the feed valve overcharging the feed-valve pipe at this time—when it is important that the supply to the governor top shall be no greater than the regulation brake-pipe pressure—the warning port of the brake valve is supplied from feed-valve pressure; this gives relief, and insures that *the excess-pressure governor head will regulate the brake-pipe pressure in* **release position** (as the brake-pipe and main-reservoir pressures are then in direct connection through the rotary valve), even though the feed valve is leaking, if not enough to be otherwise detrimental.

THE C-6 REDUCING VALVE.

Valve and Pipe Bracket Complete.

The C-6 Reducing Valve illustrated in Fig. 44 is used to regulate the pressure that is supplied to the INDEPENDENT BRAKE-VALVE and the AIR-SIGNAL SYSTEM;

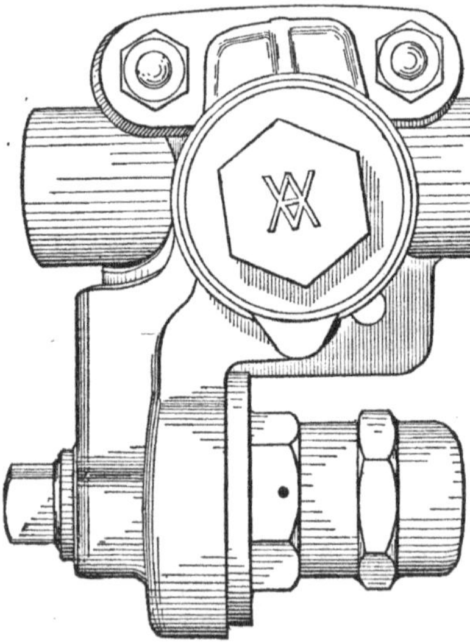

Fig. 44.—The C-6 Reducing Valve. Valve and pipe bracket complete.

it is practically the same as the feed valve just described, but without the duplex adjusting feature, being designed to reduce main-reservoir pressure to a single fixed pressure, which in this equipment is, as already stated,

45 pounds. It is, in fact, the well-known feed valve that has been used for many years in connection with the G-6 brake-valve, the only distinction being in the name; but, as here used, it is attached to a pipe bracket in the same manner as the B-6 valve. To adjust this valve, remove the cap nut on the end of the spring box; this will expose the adjusting nut by which the adjustment is effected.

The pipe bracket upon which the C-6 Reducing Valve is mounted is the same as is used with the B-6 feed valve—the crossed-passage bracket, in the No. 6 equipment, and the valve turned down in its proper position.

Fig. 45

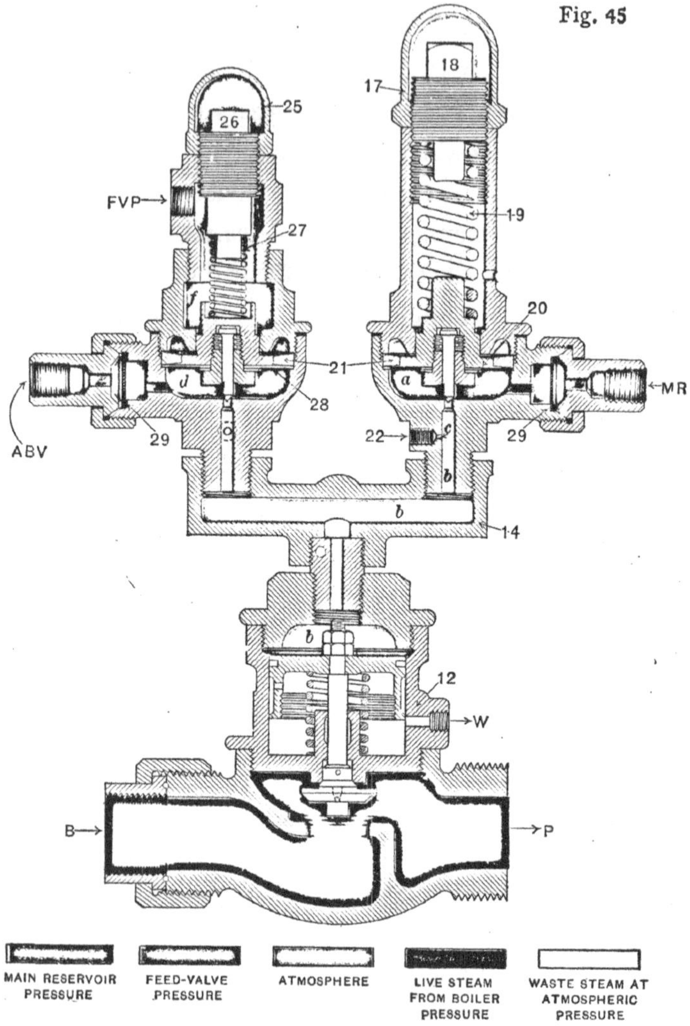

Fig. 45.—The SF-4 Pump Governor. The modified duplex pump-governor used in the No. 6 E T locomotive-brake equipment. MR—main-reservoir pipe, direct; ABV—pipe to automatic brake-valve; FVP—branch of feed valve pipe; B—steam pipe to boiler; P—connection with air pump; W—waste-pipe connection.

THE "S-F4" PUMP GOVERNOR.

The duty of any pump governor is to control the steam pressure that operates the air pump so that the pressure in the main reservoir will not exceed a given figure. With the single-top governor that figure is invariable; but with the *duplex*, or double top, S-F PUMP GOVERNOR, used in the No. 6 E-T equipment and illustrated sectionally in Fig. 45, the action of the pump is so restricted as to permit a pressure being carried in the main reservoir only 20 pounds or so in excess of that in the brake pipe while the automatic brake-valve is in **running** or **holding positions**; but when the brake-valve is moved to **application**, or **lap, positions** this governor releases the pump to increase the main-reservoir pressure to the figure fixed as the maximum. To accomplish this differential regulation only one steam valve and piston are required, and these comprise the working parts of the bottom section—being exactly similar to the corresponding section of the single-top governor; but there are two tops, or regulating sections, of the S-F

NOTE.—The figure 4 has no reference to the *type* of valve in connection with which it is used, and is only present to indicate the *size*, or, rather, the size of the steam pipe in which the governor is placed. The sizes of pipe connections are referred to in fractional *fourths of an inch*, this being understood, and only the numerator is given. A ¾-inch pump governor—that is, a governor with connections for ¾-inch steam pipe—if of the S-F type, would be referred to as "S-F 3." Hence, the "S-F4" means a one-inch governor (4-*fourths*).

PUMP GOVERNOR, and, referring to Fig. 45, the left-hand one—called the "*excess-pressure governor top*"—is adjusted to the lower pressure, and the right-hand one—the "*high-pressure top*"—is set at the pressure desired as the maximum.

The *diaphragm valves* in the two tops are exactly alike, are of the pin form, are not numbered specifically, but will be recognized as the long, central pins whose lower ends form the valves that now close ports *b*. 28 points to the diaphragm, alike in each top; 20 indicates the center piece, in particular, but generally refers to the *diaphragm-valve complete*—diaphragm, and valve, and centerpiece, etc. The *diaphragms*, 28, consist of thin discs of brass, the edges of which are tightly held by the rings 21, upon which the regulating-spring boxes seat, and are flexible toward their centers where the valve pins are located, permitting the force of the regulating springs to hold the valves down to their seats. When main-reservoir pressure in chamber *a* exerts a greater force against the under side of the diaphragm than the spring 19 does on top of it, the center of the diaphragm will be raised slightly, pulling the pin valve from its seat and permitting air pressure from chamber *a* to flow through ports *b*, *b*, to chamber *b* of the lower portion where it forces the piston downward, closing valve 5 which shuts off steam from the pump. At this time air will be heard blowing from vent port *c*. When the main-reservoir pressure in chamber *a* becomes less than the figure at which this governor top is adjusted, the superior force of regulating spring 19 bears the diaphragm center downward, reseating the pin valve, and as air supply is now cut off from chamber *b*, the pressure remaining therein quickly escapes through vent port *c;* helped by the compressed spring under the piston, steam from the boiler raises and unseats valve 5, the steam supply to the pump is resumed and main-reservoir pressure is again increased.

Boiler pressure is always under steam valve 5, and offers a high resistance to its closure; but the piston that actuates it has such a wide area exposed to air pressure that about 45 pounds per square inch on top of the piston will force the valve down against any ordinary steam pressure; therefore, it becomes the duty of the regulating portion to keep the main-reservoir pressure from flowing to chamber *b* until it has reached the desired amount. A stuffing-box and packing around the stem that connects the air piston to the steam valve, 5, would be impractical; the stem is turned to a neat fit through the wall of the steam chamber, but a certain amount of steam leaks around and past it, which, if permitted to collect, would result in boiler pressure under the piston that the air pressure above it could not overcome, and the governor would fail to regulate the action of the pump; waste port *w* allows this steam leakage to escape, however, and it should be seen that the small copper pipe connecting thereto is always open, and free from dents or sharp bends.

Reference to any of the colored Piping Diagrams will be a great help in understanding the operation of this governor, and in connection therewith note that MR is the connection of the pipe from the main reservoir; the pipe from ABV leads to the automatic brake-valve; and FVP is the connection with the branch of the feed-valve pipe.

Commonly the regulating spring of the high-pressure governor top is set at 110 pounds; its operation in pump regulation has just been explained, but although main-reservoir pressure is always present in chamber *a* —through *direct communication*—it is never high enough to unseat the diaphragm valve while the automatic brake-valve is in **release, running, or holding positions,** for then the left-hand governor top is receiving main-reservoir pressure in its diaphragm chamber, also, and it is adjusted to stop the pump when that pressure is about 90 pounds.

In the style of duplex governor used in the ordinary automatic brake equipment, the left-hand governor top is exactly like the right-hand one, with the regulating spring adjusted at 90 pounds. But in the E-T equipment, the regulating pressure exerted upon the left-hand diaphragm valve is a combination of forces

—*air*, at brake-pipe pressure, and a *spring*, having a resistance equal to the amount of excess pressure that is to be regularly carried; to secure the air pressure a branch of feed-valve pipe (see Piping Diagrams) connects with the regulating-spring box at FVP (Fig. 45); the spring 27, under the adjustment of nut 26, and much lighter than regulating spring 19 in the high-pressure top, is used, and commonly set at 20 pounds resistance. This explains why it is called the "*excess-pressure* governor top." Chamber *d* under the diaphragm has a pipe connection from ABV to the automatic brake-valve, and, as previously explained in connection with the study of the brake-valve, when in **release, running,** and **holding positions** a port in the rotary valve supplies main-reservoir pressure through this pipe to the governor (chamber *d*); when the latter pressure becomes 90 pounds, diaphragm valve 28 is unseated and the main-reservoir pressure from chamber *d* flows to the lower air-chamber *b*, forces the piston down and closes steam valve 5 as already explained.

When the automatic brake-valve is moved to **lap, service,** or **emergency positions,** the rotary valve cuts off the flow of pressure to chamber *d* of the excess-pressure governor top; the supply having ceased, vent port *c* exhausts the pressure from the piston chamber, passages and ports, *b, b, b;* the piston is relieved, steam valve 5 reopens and the pump starts up. Diaphragm valve 28 seats at almost the instant the brake-valve is lapped.

The benefits obtained from the use of the duplex pump-governor are that a moderate main-reservoir pressure can be regularly carried, reducing the wear of the pump and the rotary valve of the automatic brake-valve, and making the latter easier to operate, etc.; but during an application, the main-reservoir pressure is automatically increased to whatever figure may be considered necessary to effect the prompt release of all brakes in a train of any length, the pump only being required to raise this extra-excess pressure when it is relieved of the duty of brake-pipe supply.

The benefit from the use of brake-pipe pressure as a part of the regulating force in the governor top that normally controls the main-reservoir pressure, is in automatically maintaining a certain *excess pressure* no matter what the brake-pipe pressure may be; whatever the latter pressure is, the main-reservoir pressure will become just as much higher as the adjustment of the regulating spring in the excess-pressure governor top will permit. An engine with E-T equipment may be regularly carrying 70 pounds pressure in the brake pipe and 90 pounds in the main reservoir; if, then, it becomes necessary to operate the High-Speed Brake, the simple readjustment of the feed valve by turning the wheel handle until the handle-pin strikes the high-pressure stop will raise both pressures to the required amount. *Directly*, the readjustment of the feed valve raises the brake-pipe pressure to 110 pounds; *indirectly*, it raises main-reservoir pressure to 130 pounds, because the increase of brake-pipe pressure equally increases the air pressure upon the diaphragm of the excess-pressure governor top: 110 pounds (*brake-pipe pressure*), plus 20 pounds (*spring pressure*), equals 130 pounds, the regulating power of excess-pressure governor top, High-Speed Brake.

(Where there is a possibility of having to change from the 70-pound brake to the High-Speed Brake, the high-pressure—right-hand—governor top should not be set at less than 140 pounds, in order that it may operate at a higher figure than the High-Speed pressure adjustment of the excess-pressure top.)

Another important feature is that before commencing, and during, the descent of steep grades, this governor enables the engineer to raise and maintain the brake-pipe pressure about 20 pounds above the feed-valve regulation, merely by the use of **release position** of the automatic brake-valve, the position which should be used during such braking.

While the turning of adjustment nut 18 will increase or decrease the maximum pressure that can be obtained in the main reservoir, remember that turning adjusting nut 26 increases or decreases *the amount of excess pressure* that will be regularly carried; and that in **running** and **holding positions** of the automatic brake-valve the location of both hands on the large duplex gauge is governed by the adjustment of the feed valve; but that spring 27 in the excess-pressure governor top *keeps the two gauge hands 20 pounds apart.*

Fig. 46

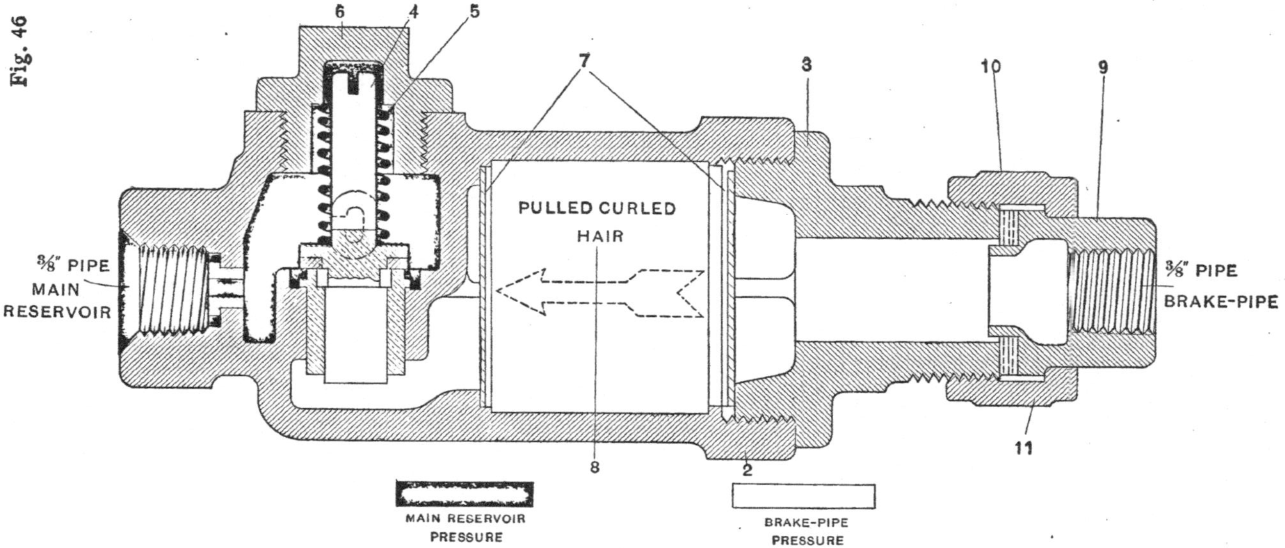

Fig. 46.—Combined Air Strainer and Check-Valve. Used as the "dead-engine feature" in all No. 6 E T equipments, and as signal-line connection with the reducing-valve pipe when the train air-signal system is supplied.

THE COMBINED AIR STRAINER AND CHECK-VALVE.

With the advent of the No. 6 E-T equipment, the COMBINED AIR STRAINER AND CHECK-VALVE illustrated in Fig. 46 is furnished as a part of the locomotive-brake equipment, whether specified by the purchaser or not; and, if the Train Air Signal is to be used, two of these are furnished. A $\frac{3}{8}$-inch cut-out cock is also supplied to be used in connection with each.

In the ordinary automatic equipment the locomotive braking power is supplied from the brake pipe (train line), and the brake on a *dead* engine is automatically operative the same as any car brake. In the E-T equipment, however, while it is automatically operated through the brake-pipe air, *locomotive braking pressure* must be taken directly from the *main reservoir*. One application of the COMBINED AIR STRAINER AND CHECK-VALVE is as the DEAD-ENGINE FEATURE, by which air from the brake pipe is supplied to the main reservoir of a dead engine, or one whose air pump is inoperative—said engine being in tow—and this is an important adjunct to the E-T locomotive-brake equipment.

Combined Air Strainer and Check-Valve

When the Train Air-Signal System is used, the Combined Air Strainer and Check-Valve forms the connection of the reducing-valve pipe to the signal pipe. The Piping Diagrams show both applications of this attachment.

As the dead-engine feature, Fig. 46 shows that the end nearest the check-valve is connected to a pipe containing main-reservoir pressure, and the opposite end to a branch of the brake pipe, the latter connection containing the cut-out cock, which should be left closed except under the conditions mentioned in which it must be opened to supply pressure from the brake pipe to the main reservoir; when open, brake-pipe air, entering as shown in the cut, passes through the disc strainers 7 and the curled-hair stuffing between the discs, lifts check-valve 4 which has been held to its seat by the strong spring 5, passes through the small choke bushing, and out to the main-reservoir connection as indicated, thus providing pressure for operating the brake on this locomotive. The Independent and Automatic Brake-Valves should be in **running position,** and the double-heading cock under the latter valve should be closed. When the tender is light of coal and water, or the locomotive boiler empty, it is commendable practice to reduce the maximum braking power of such a locomotive lower than the standard; and this can be easily and quickly done by reducing the adjustment of the safety valve on the distributing valve; however, if an engineman is in the cab this will be unnecessary, as excessive brake power can be thrown off at will by the Independent Brake-Valve.

The spring, 5, over the check-valve is made extra strong to insure the valve seating, and, although permitting ample pressure to operate the locomotive brake, keeps the main-reservoir pressure somewhat lower than that in the brake pipe, thereby reducing the chances of back leakage from the former. The small *choke port* prevents a heavy drain from the brake pipe when the uncharged main-reservoir is cut into a charged brake pipe, and operates similarly to the feed groove in a triple valve.

As the SIGNAL-LINE CONNECTION, the end nearest the check-valve connects with the branch of the main signal-pipe, and the opposite end with the reducing-valve pipe; when so used, a lighter spring is furnished for check-valve 4, and this constitutes the only difference, constructively, in the two applications of the COMBINED AIR STRAINER AND CHECK-VALVE. The *check-valve* is here necessary to prevent back flow of signal-line pressure when an independent-brake application is made, and the consequent blowing of the air whistle out of time. A *cut-out cock* should also be placed in one of the pipes connecting with this signal-system attachment, preferably in the branch pipe connecting with the reducing-valve pipe, the cut-out cock standing open normally, but necessary for the purpose of cutting out the signal line system if such should ever be required; and to facilitate the cleaning of the check valve, which should be done occasionally.

GENERAL OPERATION OF THE TRAIN AND LOCOMOTIVE BRAKES.

Before Leaving the Vicinity of the Roundhouse.

See that everything about the air-brake and air-signal (if used) systems is working properly. While the air pump should be started slowly, the air pressures should be pumped up to the limit, with the automatic and independent brake-valves in **running position.** Then note air gauges, and if either hand (or both) on the large duplex gauge does not show the desired pressure, *regulate the black hand first,* by adjusting the feed valve. Then if the red hand is not standing at the desired figure, correct that by adjusting the regulating spring of the excess-pressure governor top. Next, make a light service reduction by the automatic brake-valve, note its action, and watch the red hand rise; if it does not stop at the desired pressure, correct the adjustment of the high-pressure governor top. Return the brake-valve from lap to **holding position** and watch the red hand of the *small duplex gauge* to see that the brake-cylinder pressure will be maintained, thus making it safe to depend upon the holding power of the locomotive brake while making a running release of the train brakes on the road. Replace the brake-valve handle in **running position** and see that engine and tender brakes promptly release. Then place the independent brake-valve handle in **slow-application position** and note that red hand of small duplex gauge indicates correct adjustment of the reducing valve; push the handle to **quick-application,** and then to **release, positions** to test the return spring, and leave the valve handle in **running position.**

Before making tests, however, it is advisable to blow out the brake pipe and signal pipe at front of engine and rear of tender, by opening and closing the angle cocks and cut-out cocks a time or two.

On the Road.

Freight Service.—With long trains the best results are obtained in making service stops by *one* rather heavy reduction, as experience has proven that light initial reductions are more productive of shocks to the train when, as is most common, there is a difference of braking forces as between the forward and rear portions of the train, due to variance of piston travel, loaded cars ahead and empty ones behind, etc.

To release the train brakes, always use the **release position** of the automatic brake-valve; leave in that position until assured that all car brakes are released, then move the handle to **holding position;** when the train has stopped, or if there seems to be no reason for keeping the locomotive brake on any longer, return the handle to **running position.** After releasing the train brakes *while running,* however, and having brought the brake-valve handle back to **holding position,** leave it there until the tendency toward train stretching is past, and then release the locomotive brake by the **running position** without danger of the train parting.

After releasing the brakes of a long train, and a few seconds after the automatic brake-valve handle has been returned to **running position,** move it *again to* **release position** for about 5 seconds; this is to insure the permanent release of some of the car brakes that may have started to reapply on account of the drop of brake-pipe pressure at the forward end of the train line which always accompanies the return of the brake-valve to **running** or **holding positions** after the release of the brakes of a train of some length (this will be noticed by watching the brake-pipe gauge).

Passenger Service.—On very long passenger trains, the braking should be done the same as with freight trains, in most respects. In ordinary passenger service, however, there are certain special rules to be observed in operating the E-T brake, as follows: Two application station or service stops should invariably be made, and this method becomes imperative with the original High-Speed Brake. In releasing after the first application, the best results are secured by moving the automatic brake-valve to **running position,** and thereby releasing the locomotive brake as well as the train brakes; then, the second application will have a smooth and even effect. This is perfectly safe to do with almost any passenger train, as the brake-pipe supply ports through the H-6 brake-valve that are open in **running position** are as large as the direct-release ports, and the flow of air in the former position is only restricted by the capacity of the feed valve which in the B-6 model is ample for the purpose. Another good feature of the *running-position-release* is that the pressure in the brake pipe will not be raised above 70 pounds, and if this is followed by a slight pause of the brake-valve handle in the **lap position** there will be no brake-pipe overcharge to displace, and the brakes will immediately respond to the second application.

Use of the independent brake-valve should be avoided as far as possible while running with a passenger train, and, if used at all, the greatest care must be exercised in applying *and releasing* the locomotive brake, to avoid shocks to the train.

General Service.—To apply the brakes in emergency, move the handle of the automatic brake-valve *quickly* to **emergency position,** and leave it there until the train stops or the danger is past.

When the train and locomotive brakes are applied and it becomes necessary to release the locomotive brake only, it is accomplished by holding the independent brake-valve in the **release position** until the brake-cylinder pressure is partially or completely exhausted, as may be desired, then bring the handle to **running position** and leave it there. Or, with the train and locomotive brakes applied, if it is desired to increase the braking power of the locomotive, use the **application position** of the independent brake-valve and return it to **running position.**

When the automatic brake-valve is in **running position** an application of the locomotive brake by the independent brake-valve can always be released by simply returning the handle of the latter valve to **running position.**

Use the independent brake-valve exclusively when without a train; with a train, use it only when absolutely necessary, and then with the greatest care.

Before leaving the engine while doing work about it, or when it is standing at a coal chute or water plug, on the turntable, etc., always leave the independent brake-valve handle in **application position.**

In case of train parting, or other causes of automatic applications of the brakes, such as a burst hose, use of the conductor's valve, etc., place the handle of the automatic brake-valve in the **lap position:** this to save the main-reservoir air from blowing away, and to assist the application of the brakes (from an application of this kind the locomotive brake will hold with full power whether the brake-valve is lapped or not; whereas with the No. 5 equipment the locomotive brake could not be applied automatically while the brake-valve was in *running position*).

In heavy grade service, **release position** of the automatic brake-valve should always be used. In order to prevent overheating of driving-wheel tires, and to assist the pressure-retaining valves in holding the train while the auxiliary reservoirs are being recharged, it is recommended to work the independent locomotive-brake and the train brakes alternately; this may be done by holding the independent brake-valve in **release position** while the train brakes are being applied by the automatic brake-valve, and applying the independent brake just before releasing the train brakes.

The independent brake will hold a locomotive with leaky throttle valve, or quite a heavy train on a fairly steep grade after having stopped, if it is solidly applied before the train brakes are released. But remember, always, that when the independent brake is to be relied upon, absolutely, the independent brake-valve *must be left in* **application position,** and *not be moved back* to the position of **lap.**

When there are two or more locomotives in a train, the double-heading cock must be closed and the handle of the automatic brake-valve carried in **running posi-**

tion, on each engine except the one from which the brakes are operated.

Whenever an application of the train and locomotive brakes has been made on a double- or triple-headed train, while running, the automatic brake-valve on each "*cut-out*" engine in the train should be placed in **Holding Position,** in expectation that a slow-speed, running release may be made; and if it should, the brakes of the "*cut-out*" locomotives will be retained, to help hold the forward portion of the train from surging ahead and possibly breaking in two; after the train brakes have all fully released, the automatic brake-valve handles on the secondary locomotives should be returned to **Running Position,** in which the locomotive brakes will release.

When coupling to a train whose brake pipe and auxiliary reservoirs are empty or at a low pressure, if the automatic brake-valve handle is permitted to remain in **running position** the air pump will stop working, and will not restart until the fall of main-reservoir pressure and the rise of brake-pipe pressure has brought the two pointers on the large duplex gauge within less than 20 pounds of each other. This does not indicate a defective condition anywhere, and is perfectly natural to the E-T equipment in general. The proper procedure is always to place the automatic brake-valve in **release position** at such a time, and do not return it to **running position** until the two gauge hands are within 15 pounds of each other—never permitting the black hand to rise above 70 pounds, permanently, however.

ON ARRIVAL AT ROUNDHOUSE AT FINISH OF TRIP.

Reporting Air-Brake Repair Work.—The best policy is for an engineer himself to inspect, clean and oil the equalizing-discharge piston and rotary valve of the automatic brake-valve, and the rotary valve of the independent brake-valve; to attend to any necessary adjustment of the feed valve and reducing valve, and the regulation of the pump governors. In respect to the latter, there is a fine strainer in each main-reservoir pressure pipe-connection to the governor tops which should be cleaned, and the pipes blown out occasionally.

If the return spring in the independent brake-valve gets broken, do not fail to insist on having a new spring put in at once. Don't make a single trip without this spring being in good condition, or else the handle may be forgotten in the **release position**—warning port gummed up—and in case a quick stop is seriously necessary, the failure of the locomotive brake to apply (which will be the result) may be disastrous.

Become well enough informed on the E-T brake to be able to make accurate *and intelligent* reports of necessary work. Make no such indefinite reports as "*brake valve not working properly*"; state the trouble with the brake-valve, and you won't have any complaint to make against the repair men, nor trouble in handling the train next trip on account of *brake-valve not working properly.*

Don't allow your driver- and tender-brake piston travel to become too long, and don't permit brake-cylinder leakage to exist, just because these things haven't interfered so far with the power and holding effect of your locomotive brake; don't let the driver-brake pistons run out much more than 4 inches, the tender-brake piston 7 inches, and the locomotive truck-brake piston 6 inches. If you have to shorten the piston travel yourself, take up the slack of the driver-brake rigging by means of the screw take-up arrangement in the brake rods near the cylinder-lever connections, as nearly equal on both sides as possible, and so that the shoes will hang as close to the wheels as possible without contact with the tires. The same applies to the tender brake, taking up its slack by means of the dead truck-levers, an equal amount on each truck, but do not let its piston travel be shortened to less than 5 inches.

Testing for Leaks in the No. 6 *Equipment.*—This concerns the SHOP MAN as well as the ENGINEMAN. Any usual reason that may be given to explain certain leakage from one pipe or air chamber to another can be counterfeited by leaks between ports in the *distributing-valve gasket,* although it is unusual for this to happen. The gasket is shown in Fig. 47; the different pressures that are ported through it are named thereon, and it can be seen that any number of pressure combinations *may* occur from intercommunication of the ports. To avoid possibility of such trouble, before the valve section is re-attached to the reservoir section after removal it should be seen that the gasket is in perfect condition, that no parts are torn from it and sticking to the faces of the sections of the distributing valve; then rub *dry* graphite on both sides of the gasket, and after the two sections are together tighten the bolts alternately and gradually, working around until all are perfectly tight.

The same thing applies to gaskets 18 and 19 of the automatic brake-valve, and gasket 25 of the independent brake-valve, to a certain extent. See that the long bolts through the brake-valves are always perfectly tight, and if either valve is placed so close to the boiler that the heat will harden and crack the gaskets, have the brake-valve relocated in a better place.

In erecting the piping, it should be seen that the pipe to the black hand of the small duplex gauge is connected to a tee in the brake pipe *below the double-heading cock,* so that brake-pipe pressure will be indicated when the brake-valve is cut out.

Have all pipes secure against vibration. Have the feed valve and reducing valve bolted firmly to an iron support; don't compel the pipes to support themselves and the heavy valves, too.

If anything about the tender brake becomes defective, necessitating cutting it out, remember the cut-out cock

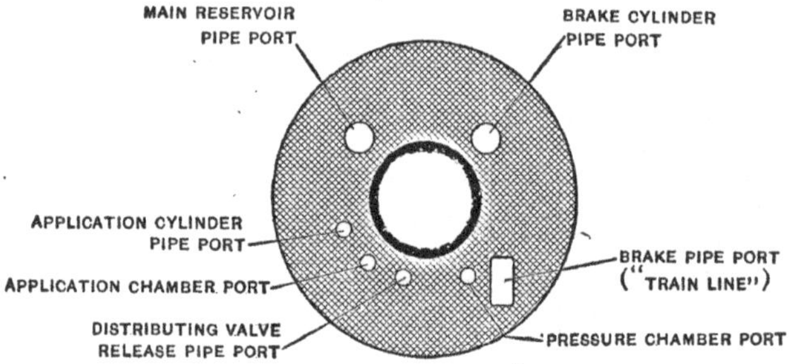

FIG. 47.—Distributing-Valve Gasket. No. 6 distributing valve. Used between valve section and double-chamber reservoir.

for that purpose *is not on the tender,* but is located on the engine, in the branch of the brake-cylinder pipe that leads to the tender brake, and close to the hose connection with tender.

Regular signal hose, one pair complete, are used in the brake-cylinder pipe line between engine and tender. When the engine-truck brake is part of the equipment, the pipe line to truck brake terminates in a single-hose connection with the brake cylinder, the ends of this hose being fitted with union and nipple, respectively.

See that the pipe conveying main-reservoir air to the high-pressure governor top is connected to the *main-reservoir cut-out cock*, and that the cock is so put up that the governor-pipe connection will be on the main-reservoir side. If the cut-out cock is located on the opposite side of the engine from the pump governor, connect this governor pipe to a branch of the main-reservoir pipe that can not be cut out; this to insure the pump always being under the control of the governor, for when the main-reservoir cut-out cock is closed the pressure will be exhausted from all other parts of the locomotive air-brake equipment.

BROKEN OR LEAKING PIPES.

Main Brake Pipe ("train line") *Under Tender.*—When broken or badly leaking, dispense with the use of the train air signal, and use the main signal pipe under the tender in place of the defective brake pipe. Every engine should carry two "*combination hose*" complete—very short hose with brake couplings on one end and signal couplings on the other end—for use in such cases, and also when the brake-pipe rupture occurs under any car in a passenger train and it may be undesirable to switch the car to the rear of the train. Place the automatic brake-valve on **lap,** and close the cut-out cock in the signal line supply-pipe near the combined air strainer and check-valve; close both angle and cut-out cocks between rear of tender and first car; separate the brake and signal hose connections between engine and tender, and tender and car; recouple the brake hose from engine to signal hose on tender, and signal hose from rear of tender to brake hose on first car; open cut-out cock at rear of tender, and angle cock at head end of the car; place automatic brake-valve first in **release,** then in **running position,** and you are ready to proceed.

Equalizing-Reservoir Pipe.—With this pipe broken, stop the flow of air from the rupture by putting a blind gasket in the union of the tee-fitting by which it connects with the automatic brake-valve; plug the service-exhaust opening under the brake-valve, and proceed —carrying the brake-valve in **running position** as usual. In making stops, no results will follow the movement of the brake-valve handle to the **service-application position,** except that the black hand of the large duplex gauge will instantly drop to zero. Do your service braking by jumping your automatic brake-valve handle over the service-stop shoulder, cutting-in to the **emergency position** just far enough to draw off brake-pipe air *directly,* but comparatively lightly, through the "big hole" of the brake-valve. The application can be graduated very nicely, if care is used, noting the amount of brake-pipe reduction that is being made *by the black hand of the small duplex gauge.* Return the handle to

lap position rather slowly, or some of the forward brakes will "kick off"—release themselves.

Main-Reservoir Supply Pipe to Distributing Valve.—With this pipe broken beyond repair, the *locomotive brake* is inoperative by either brake valve. If broken between the cut-out cock and distributing valve, simply close the cock; if the rupture is between the cut-out cock and main reservoir, plug, or place a blind gasket in the pipe toward the latter pressure: or, if conditions permit, remove the pieces of broken pipe and re-attach the closed cut-out cock, so as to shut off the escape of pressure. Proceed, bearing in mind the absence of locomotive braking-power.

Brake-Pipe Branch to Distributing Valve.—Judging from results on different roads, this is most frequently broken of any part of the E-T-equipment piping. When it becomes ruptured the locomotive and train brakes will apply; **lap** the automatic brake-valve, permitting the brakes to stop the train; then plug the broken pipe toward the brake-pipe pressure, release train brakes, and proceed. The locomotive brake will not then operate from an automatic application, but can be applied by the independent brake-valve, although the **release position** must be used to release it.

Brake-Cylinder Pipe From Distributing Valve.—A cracked joint occurring at any point in this pipe line while on the road, if not opened too wide, may not weaken the pressure in any of the locomotive brake-cylinders, but there will be a continuous blow of air from the point of rupture when the locomotive brake is applied. If the pipe is nearly or quite broken off between the distributing valve and brake-cylinder cut-out cocks, at the first stopping point close the cock in the main-reservoir supply pipe to distributing valve, and proceed without the locomotive brake. A break in this pipe line between either brake cylinder and its individual cut-out cock, will only deprive the locomotive of the power of that one brake cylinder, for, on account of the small choke-fittings in the brake-cylinder cut-out cocks, the pressure can not blow away from the other brake cylinders of the locomotive as fast as the application valve of the distributing valve can supply it. At the first stop, close the cut-out cock in the branch pipe that is affected.

Application-Cylinder Pipe.—As the result of *a leak* in this pipe, an automatic or independent application might or might not set the locomotive brake, depending upon the extent of leakage, but after placing the brake valve in **lap position,** the locomotive brake would release itself—if it set at all. If not possible to remedy the defect, plug the application-cylinder pipe toward the distributing valve; the locomotive brake can then be applied as usual by the automatic brake-valve, and released by that valve in **running position,** but the independent brake-valve will be powerless to apply or release it.

Distributing-Valve Release Pipe.—The breakage of this pipe need not cause any delay, nor will it affect the braking power of the locomotive *during an application;* but when the automatic brake-valve is placed in **release or holding positions** the usual effect is not wholly obtainable, the locomotive brake will release at once; and with a long train the train brakes should be held on until coming to a dead stop after all applications while running. The locomotive brake can be applied by the independent brake-valve, but will release if that brake-valve is placed in the **lap position.**

Feed-Valve Pipe Branch to Excess-Pressure Governor Top.—If it breaks off, plug it toward the feed-valve pipe; the excess-pressure governor top will not then permit the pump to work when the main-reservoir pressure is as high as 45 pounds, while the automatic brake-valve is in **release, running,** or **holding positions;** to remedy this, place a blind gasket in the pipe leading from the automatic brake-valve to the chamber under the diaphragm of the excess-pressure top, which completely cuts that governor top out of service; the pump will then be controlled solely by the high-pressure governor top, and will regularly maintain the main-reservoir pressure at the maximum figure.

Main-Reservoir Pressure Pipe from Automatic Brake-

Valve to Excess-Pressure Governor Top.—If this pipe breaks *en route,* place the automatic brake-valve on **lap,** and quickly plug the pipe toward the brake valve; then move the brake valve into **release position** for about three seconds, and back to **running position.** Results will be the same, then, as after remedying for the last mentioned defect, the high-pressure governor top regulating the pump's action.

Main-Reservoir Pressure Pipe (direct) *to High-Pressure Governor Top.*—Plug this pipe, when broken, toward the main reservoir, and go right along; but as the excess-pressure governor top then exercises the only automatic control of the pump—and it only when the automatic brake-valve is in **release, running,** and **holding positions,**—when the brake-valve is in **lap, service, or emergency positions,** the gauge hand indicating main-reservoir pressure should be watched, and if it begins to get too high the steam should be eased off from the pump by closing the throttle at the boiler, as far as may be necessary.

Reducing-Valve Pipe.—Rupture of this pipe has a farther-reaching effect than merely in cutting off the supply of pressure to the independent brake-valve, and possibly the air-signal system. Stop the flow of air from the broken pipe, by turning the adjusting nut so as to loosen the regulating spring of the reducing valve, until the blow of air ceases. Letting it go at that, it would also be impossible to secure an *automatic* application of the locomotive brake; and the further remedy is to plug the broken pipe *toward the independent brake-valve,* and plug the exhaust port in the bottom of the brake-valve. The independent brake-valve is then valueless, except as a fixture, the handle of which must not be moved from **running position;** but the operation and holding power of the automatic brake of the locomotive will not be affected.

Pilot Section of Brake Pipe.—An angle cock is not generally used at the pilot end of the brake pipe, an angle fitting taking its place, and a cut-out cock is placed in the brake pipe back of the engine cylinders. Therefore, when coupling to a train in front of the locomotive, it is not uncommon to find that the pilot section of this pipe—forward of the cut-out cock—is broken. In such case, when it becomes necessary to couple the air from the locomotive to the train ahead—and there are no cars behind—use a "*combination hose*" to connect the brake hose and signal hose together *at rear of tender,* and open their angle and cut-out cocks; use the other "combination hose" at the pilot, to couple the signal hose from the engine to the brake hose of the car; open the angle and cut-out cocks in the connected line at the pilot, and close the cut-out cock in the pressure-supply pipe to the air-signal line, and the locomotive and train brakes are then under regular control by the operation of the automatic brake-valve.

TESTING.

For use in making *roundhouse tests* of the E-T equipment, a test gauge with a discharge cock (use a ¼-inch release cock), a 1-inch brake-coupling, and a signal coupling, the couplings with threaded shanks connecting to two of the openings of a ¾-inch 3-way cock, and the gauge at the other opening. First, connect this testing instrument to the brake hose-coupling at rear of tender, or at the pilot, with the automatic brake-valve in **release position;** open the angle cock, and set the 3-way cock so the test gauge will receive the pressure, and close the small discharge-cock on the testing instrument; then see that the red hand of the large duplex gauge, and the black hands of both air gauges, register together, and to correspond with the test gauge; or note any errors—a variance of two pounds demanding correction of the gauge. Have some one then to make a service reduction, reducing the brake-pipe pressure to about 10 pounds below normal (to 60 pounds), and then to replace the automatic brake-valve handle in **running position**—this to test the sensitiveness and the adjustment of the feed valve. Note the rise of the hand on the test gauge, which should stop at 70 pounds. Then open the small release cock on the testing instrument, and consider the action of the hand on the test gauge; if it fluctuates, falling as much as 2 pounds or more, or shows a slow overcharge of the brake pipe, the feed valve should be cleaned and oiled.

Next, connect the testing instrument—release cock closed—with the signal-hose coupling at front or rear of the locomotive; set the 3-way cock to receive that pressure, and open the signal line cut-out cock; place the independent brake-valve in **application position.** and compare the indications of the red hand of the small duplex gauge with the hand on the test gauge; this also shows the adjustment of the reducing valve, and to test its sensitiveness open the small discharge cock on the testing instrument and watch for fluctuations, same as in case of the feed valve.

To test the pump governor: with automatic brake-valve in **running position** note that main-reservoir pressure is registered 20 pounds (or the amount of excess pressure to be regularly carried) higher than brake-pipe pressure, correcting any error in this by readjustment of the regulating spring of the excess-pressure governor top. Then place the brake-valve handle in the **lap position,** in which the red hand of the large duplex gauge should register the higher, or maximum, main-reservoir pressure that is standard for the class of engine.

Testing the *automatic brake-valve*, first make a 5-pound service reduction and **lap** the valve; if the black hand of the large air-gauge continues to fall, slowly, and there is a blow of air from the equalizing-discharge port under the brake valve, there is a leak of equalizing-reservoir pressure, which must be stopped before proceeding further. Afterward, make a 20-pound service reduction of brake-pipe pressure, **lap** the brake-valve, and close the double-heading cock beneath it; if, then, the black hand of the large air-gauge gradually rises, it indicates a leaky rotary valve, or a leakage in the middle gasket, 18, of the automatic brake-valve.

Test locomotive brake-pipe leakage by making a 7-pound service reduction, **lapping** the brake-valve and closing the double-heading cock beneath it; then time the fall of brake-pipe pressure as indicated by the black hand of the small duplex gauge, and which should not exceed 5 pounds in one minute.

To test for locomotive brake-cylinder leakage, make a full-on application by the independent brake-valve, close the cut-out cock in the main-reservoir supply pipe to distributing valve, and note the brake-cylinder leakage as indicated by the red hand of the small duplex gauge. To find out which of the brake cylinders are leaking, shut off each one in turn by closing its individual cut-out cock and timing the fall of the cylinder-pressure gauge hand; when the fall of pressure is materially lessened with a brake cylinder cut-out, its leaky condition is evident.

After an application of the locomotive brake, if, in slowly releasing it by the independent brake-valve, the red hand of the small duplex gauge falls as would be expected, but at no time during the release does any brake-cylinder pressure escape from its natural exhaust opening in the front of the distributing valve, the cause is due to a bad condition of the application portion of the distributing valve; the leather packing of the application piston may be in good condition, but the packing ring is not; the ring may be worn, or broken, or stuck tight in its groove by gummy dirt and will not expand to make a tight joint. This is probably aggravated by an unnatural resistance of the application and exhaust slide valves, from being dirty and lacking lubrication. Some insistence is often necessary to get inspectors to clean and oil the application piston and its connecting valves, on account of the trouble in getting it out, as, besides taking off the application-cylinder cover, the top cover over the application valve must also be removed, involving the taking out of the many little screws that hold it; the application valve must be lifted off, and the pin that operates it pulled out of the application-piston spindle, before the piston can be removed. But, if any part of the E-T equipment should be slighted in care, *it must not be the distributing valve*—the fundamental "hub" of the locomotive braking system.

Directions for testing the different parts of the No. 6 E-T equipment for the many *possible* defects could be continued almost indefinitely, but to give them all in detail would imply that the reader, or student, is not capable of understanding when and how a certain part is working defectively *after he has learned how it should work when it is operating correctly*, and if the implication should be correct the details would be an overtax on his memory. The main essentials in roundhouse testing of this equipment are given above; but the chief air-brake inspector, the roundhouse foreman, and the back shop air-brake repair men, should become so thoroughly acquainted with the No. 6 E-T equipment by a complete understanding of the subject matter of this book, that they will be able to detect the many possible irregularities of the equipment; in fact, to so educate them—and locomotive enginemen—is the object of this work.

THE NO. 5 E-T LOCOMOTIVE-BRAKE EQUIPMENT.

Its General Construction.

Probably the larger number of E-T-equipped locomotives at the present time have the former, No. 5 STYLE, which was discontinued with the advent of the No. 6, and all locomotives recently built have the later, improved type, as all will have in the future. As a fact, there is but very little difference between the two styles of this equipment, and if either one is well understood it will only require a few words of explanation to make the other style equally clear.

Fig. 48 shows the No. 5 E-T Equipment; and the only difference between this plate and a diagram of the No. 6 style is in the *two small copper pipes* connecting with the left side of the distributing valve—the only ones shown in colors in this cut, as all other parts are exactly similar in appearance, pressures contained, and their duties, to the corresponding parts now well understood in the No. 6 equipment. The *orange-colored* pipe—lower connection on the left side of distributing valve—performs, to a certain limit, the duties of both of the "two little copper pipes" of the No. 6 equipment; while the *blue-colored* pipe plays no part whatever in any of the phases of brake operation (except an *undesirable*

Fig. 48

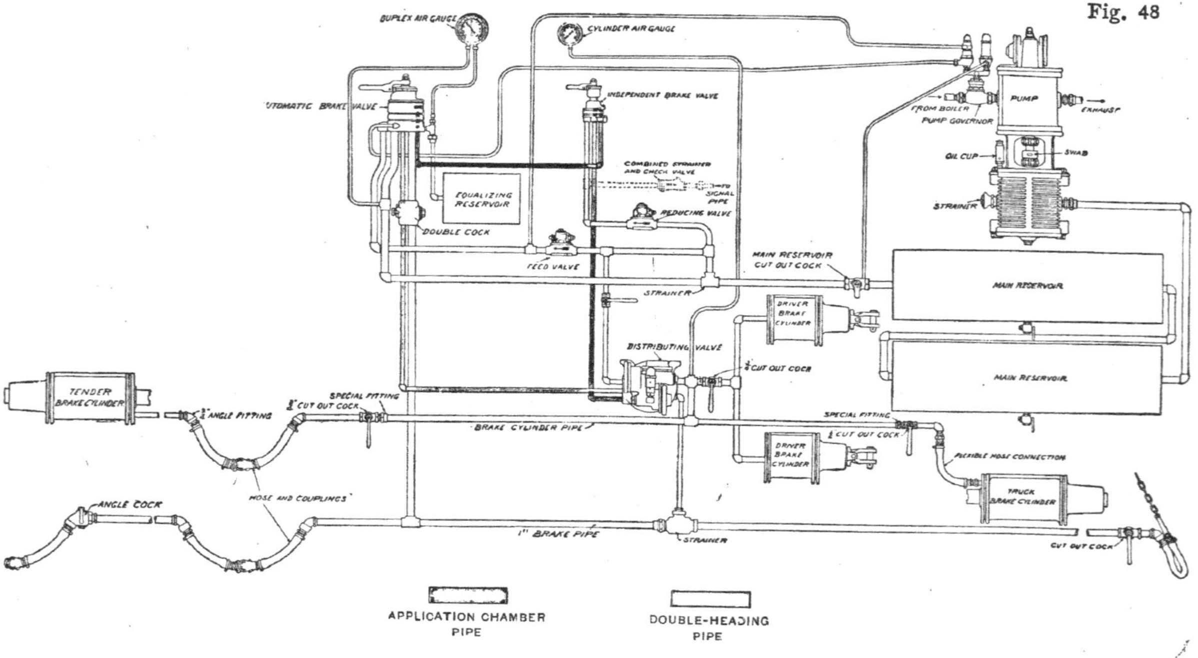

Fig. 48.—Diagram of the No. 5 E T Locomotive-Brake Equipment. Colors showing the only apparent difference from the improved No. 6 equipment.

part, when leaking), unless the locomotive is placed in a train of which the brakes are operated from another engine; hence the name of the latter pipe, as given in the Color Key—the "*Double-Heading Pipe.*"

The *orange*-colored *Application-Chamber Pipe* therefore becomes a most important part of this equipment; the application chamber and application cylinder are in permanent communication with each other in the No. 5 distributing valve, and from those combined chambers the *application-chamber pipe* leads to the atmosphere, *via* the independent brake-valve, at the automatic brake-valve—when both brake-valves are in **Running Position;** through this pipe the locomotive brake is released when the automatic brake-valve is returned to **Running Position** after an application, and applied or released by the independent brake-valve. It will be observed that when the automatic and independent brake-valves are placed in **Release Position,** pressure that may have been contained in the application chamber and application cylinder will be exhausted, and the locomotive brake released regardless of whether the equalizing, or *triple-valve*, portion of the distributing valve is *in* **release position** *or not;* and from this it follows that when an automatic application is made from the train—by use of the conductor's valve, an angle cock, or from the train parting—there is a special reason for placing the automatic brake-valve handle in the **Lap Position,** for otherwise the locomotive brake will not hold; if this duty should be delayed, and the brake-valve handle later be placed in the **Lap Position,** the locomotive braking pressure will be built up, however, by the *maintaining pressure* which in this style of equipment originates in, and is supplied by, the distributing valve.

The *blue*-colored or *Double-Heading Pipe* leads from the exhaust port of the equalizing, or *triple-valve* (lower), portion of the distributing valve to the double-heading cock under the automatic brake-valve, and under ordinary conditions is blanked by the latter. The double-heading cock has two ports through it, the ports in the cock key being at right angles to each other; when the cock is "open," brake-pipe pressure flows through it, and the port connecting with the double-heading pipe is closed; when, as on the second engine in double-heading, the double-heading cock is "closed," while it does cut off connection between the automatic brake-valve and the brake pipe, the smaller port in the cock key is then open, connecting the section of the *blue* pipe that comes from the distributing valve with the upper section of that pipe-line leading to the automatic brake-valve; but here the pipe is again blanked, until the brake-valve handle is placed in **Lap Position,** in which, through a port in the rotary valve, the double-heading pipe line finds an exit to the atmosphere at the large, emergency-exhaust opening. Outside of the two colored pipes, as noted, the whole No. 5 equipment is about the same as the No. 6.

In the distributing valve, the upper, or *application, portion* is exactly the same; and the lower, or *equalizing, portion* only differs in a slight variation of the ports in the equalizing slide valve, and the absence of the graduating spring that is supplied in the No. 6 distributing valve. The safety valve on the No. 5 distributing valve is set at 53 pounds, instead of 68 pounds as in the No. 6 equipment. The small air gauge is of the single-pointer style, the one (black) hand registering locomotive brake-cylinder pressure. Refer to the "No. 5 Distributing Valve," the "H-5 Automatic Brake-Valve," the "No. 5," or "S-F," "Independent Brake-Valve," the "B-4 Feed Valve," the "B-3 Reducing Valve," and the "S-F4 Pump Governor," in this style of equipment.

Handle Positions of the Engineer's Brake-Valves, No. 5 E-T Equipment.

The Automatic and Independent Brake-Valves have the same number of operating positions, each, in the No. 5 as in the No. 6 equipment, and their action is so nearly the same that if an engineer was used to either style, he could operate the other without any special instructions, *in ordinary service;* but there is some difference in the work performed by the several parts of the equipment, which will be briefly mentioned. It will be assumed that an automatic application had been made:

THE H-5 AUTOMATIC BRAKE-VALVE.

In **Release Position** of the handle, the results obtained are precisely the same as explained in reference to the H-6 valve; main-reservoir pressure is supplied directly to the brake pipe, releasing the train brakes and moving the equalizing portion of the distributing valve to **release position,** which, in addition to permitting the recharge of the pressure chamber, connects the application-chamber pressure with the blanked double-heading pipe. Application-chamber air fills the application-chamber pipe from the distributing valve, through the rotary of the independent brake-valve, to the automatic brake-valve, where it is blanked by the rotary valve. Main-reservoir pressure is being supplied from the automatic brake-valve to the chamber under the diaphragm of the excess-pressure governor top, controlling the pump at the minimum m.-r. pressure. Air to the *warning port* is from main-reservoir pressure direct.

In **Running Position** the direct supply of main-reservoir pressure to the brake pipe is cut off, and the latter receives its pressure from the 70-pound feed-valve pipe, as usual in the E-T equipment. A port in the rotary valve now opens the terminus of the application-chamber pipe to the atmosphere, and the locomotive brake releases (the only position of the brake-valve in which it does discharge that pressure). The excess-pressure governor top still receives main-reservoir pressure from the brake valve.

In the **Holding Position,** feed-valve pressure supply to brake pipe is continued, but the application-chamber pipe is again blanked; and if the brake-valve handle had been drawn quickly to this position from that of **Release,** the effect would be the same as in **Running Position,** except that the locomotive brake would remain applied. After using this position as long as may be necessary, release the locomotive brake by returning the brake-valve handle to **Running Position,** and leave it there. Pump control is the same as in **Running Position.**

Lap Position.—In the three first positions of the brake-valve, just mentioned, chamber D and its connected equalizing reservoir received the same pressure that was supplied to the brake pipe; but in **Lap, Service,** and **Emergency Positions,** chamber D is cut off from the air supply. Also, in **Lap Position** the feed-valve pressure no longer flows to the brake pipe, and all separable communications in the rotary valve and seat are blanked—except one connection that is made in this position, only:—the terminus of the double-heading pipe is opened to the atmosphere through a port in the rotary valve, but which ordinarily has no effect, however, as this pipe is closed at another point —at the double-heading cock. In this, and the following **application positions,** the supply of main-reservoir pressure through the rotary valve of the automatic brake-valve to the excess-pressure top of the pump governor is cut off, and the pump is then solely under the control of the high-pressure governor top.

(**Lap Position** is the carrying position for the automatic brake-valve on all engines that may be in a train, *except the leading one, or the engine that is to operate the brakes;* and on such secondary engines the double-heading cocks should be closed—to brake-pipe air,—and this completes the opening of the double-heading pipe to the atmosphere; having the effect of converting the equalizing portion of the distributing valve into *an actual triple valve* with open exhaust port, that can be automatically operated by the leading engineer in both application and complete-release movements.)

Service-Application Position.—The movement to this position does not change the lapped condition of the rotary-valve ports, except that the terminus of the double-heading pipe is closed, and another port is now opened: the small, preliminary-exhaust port is open, discharging equalizing-reservoir pressure from chamber D, which has the already well-known effect of initiating an equal pressure reduction of brake-pipe air through the service exhaust port. A return of the handle to **Lap Position** follows, when the equalizing-reservoir pressure has been reduced the desired amount.

Emergency-Application Position.—Like the position for **service application,** that of **Emergency** has the same effect on the train brakes as similar positions of the H-6, or any other standard Westinghouse brake-valve. In **Emergency Position** a large port in the rotary valve connects the brake-pipe pressure with an equally large port in the rotary-valve seat that opens to the atmosphere, causing such a quick and heavy reduction as to apply all brakes at emergency, or *quick action.*

And, as in the No. 6 equipment, an emergency application produces a higher application-cylinder pressure than is obtained at a full service reduction; this pressure increase was shown to be obtained in the improved, No. 6 brake, by the cutting off of the application chamber at emergency action of the distributing-valve parts, and causing the *short equalizing* of the pressure-chamber air with the small, application cylinder; in the No. 5 equipment, the application chamber remains in permanent communication with the application-cylinder in all phases of distributing-valve action, and its increased pressure is only obtained by placing the automatic brake-valve in **Emergency-Application Position,** in which position a port in the rotary valve opens communication between chamber D and the application-chamber pipe; and the addition of the volume and unreduced pressure of the equalizing reservoir to that of the pressure chamber builds up a pressure to act upon the application piston of 60 pounds; whereas a full service application can only equalize the pressure chamber, application chamber and application cylinder, at 50 pounds pressure.

Maintaining pressure, supplied at a certain time to the application cylinder of the distributing valve, and as understood in the No. 6 equipment, is also a factor in the No. 5 brake operation, but there is an important difference in its origin—an improvement in the No. 6, decidedly for the better. In the No. 5 brake, the feed of maintaining pressure is not dependent upon the placing of the automatic brake-valve in **Emergency Position**; it does not come through the brake-valve at all, originating in the distributing valve; when the equalizing slide valve is drawn to the limit of its application movement, a port through it connects the main-reservoir pressure (which is always present in the distributing valve) with a port in the valve seat that leads to the application cylinder *via* the pressure chamber and application chamber; it will be seen from this that a full service application will produce this "*straight-air*" supply, which is not at all desirable.

THE S-F (No. 5) INDEPENDENT BRAKE-VALVE.

The OPERATIVE POSITIONS of the S-F INDEPENDENT BRAKE-VALVE handle are the same as with the S-6 valve; the positions of **Slow Application** and **Quick Application** are closer together in the No. 5 valve, however, and the shoulder on the quadrant that catches the handle bolt in the former position is not as pronounced. The return spring only operates to move the handle from **Release** to **Running Position**; and there is no *warning port* to attract attention in the **Release Position** of the No. 5 (S-F) VALVE. The duties of these two styles of INDEPENDENT BRAKE-VALVE in their five operative positions are just the same, in each, and have been fully described in explanation of the S-6 brake-valve.

GENERAL OPERATION OF THE No. 5 E-T EQUIPMENT.

AS IT MAY DIFFER FROM THE IMPROVED, No. 6 STYLE.

In general operation it may be said that there is no difference in the instructions as between the Nos. 5 and 6 styles of E-T equipment, with the single exception of double- or triple-heading, and then so far only as concerns the "*cut out*" engines; on each engine in the train except the one from which the train and locomotive brakes are operated, the independent brake-valve should, as usual, be carried in **Running Position** and the automatic brake-valve must be placed in the **Lap Position,** with the double-heading cock beneath it "closed" to brake-pipe pressure. Under such arrangement, in order to apply the independent locomotive-brake the automatic brake-valve handle must first be placed in **Holding Position;** to graduate the application, the independent brake-valve should then be placed in **Slow Application Position** and returned to **Running Position.** To release the locomotive brake afterward, simply return the automatic brake-valve handle to **Lap Position.**

After an automatic train- and locomotive-brake application from the operating engine, if it is desired to retain the driver and tender brakes on a "*cut-out*" engine while the train brakes are being released, the automatic brake-valve should be temporarily placed in **Holding Position** until it is desired to release the locomotive brake, which will be done by returning the handle to **Running Position.**

The DEAD-ENGINE FEATURE is not included in the No. 5 E-T brake equipment, and should be specially ordered, as its application is strongly recommended.

In the earlier furnishings of the No. 5 equipment, the branch of the feed-valve pipe to the excess-pressure top of the pump governor was connected by a union to the top of the regulating-spring case, and a stop cock was placed in the pipe near the governor; in order to correct or change the adjustment, with this arrangement, it is necessary to first close the stop cock, then disconnect the pipe from the spring case, remove the cap nut, and turn the regulating nut in the governor top *by guess;* after which the cap nut is screwed on, the pipe reconnected, and the stop cock reopened; by consulting the red hand of the large duplex gauge, it can be ascertained how near to the desired figure the adjustment has been made; and usually the same procedure will have to be gone through several times before the adjustment is correct. On most roads this older style of excess-pressure governor top is being replaced by the newer style, explained in connection with the No. 6 equipment, but which is now the standard for both styles of the E-T equipment. If the stop cock in the feed-valve pipe branch to the older-style governor top should get accidentally closed (as it often does), not more than 45 pounds' pressure can be pumped up in the main reservoir while the automatic brake-valve handle is in **Release, Running,** or **Holding Positions**; but in the positions of **Lap, Service,** or **Emergency Application,** the main-reservoir pressure will be pumped up to the maximum figure.

LEAKING OR BROKEN PIPES IN THE NO. 5 EQUIPMENT.

With the exception of the two small, copper pipes known as the *application-chamber pipe,* and the *double-heading pipe,* shown in colors in Fig. 48, the results from pipe leakage, or breakage, and the remedies therefor, are just the same in the No. 5 equipment as already explained in reference to the No. 6-equipment piping—with one exception: If the BRAKE-PIPE BRANCH TO DISTRIBUTING VALVE in the No. 5 equipment becomes broken, it is usually impossible to go right on after simply plugging the ruptured pipe, without the locomotive brake sticking—brake shoes rubbing the wheels, and a distressing blow of air from the main exhaust port of the automatic brake-valve; and at the first light application by either the automatic or independent brake-valve, the application-chamber pressure will run clear up to 53 pounds and "pop" at the safety valve. To avoid this trouble, when the brake-pipe branch gets broken, close the cut-out cock in the main-reservoir supply pipe to distributing valve, and proceed (after plugging the broken pipe), without the use of the locomotive brake in any kind of application whatever.

If the APPLICATION-CHAMBER PIPE (*orange*-colored, see Fig. 48) starts leaking at any point between the distributing valve and the independent brake-valve, the locomotive brake will not hold—probably not set at all—as the result of any brake-pipe pressure reduction; if the leak is not too bad, an application by the independent brake-valve will cause the locomotive brake to hold as long as it remains in the **Quick-Application Position,** but the brake will release as soon as the brake valve is lapped. Leakage only from the section of this pipe connecting the independent and automatic brake-valves will have no effect whatever upon an application by the independent brake-valve, but the locomotive brake will not apply from any reduction of brake-pipe pressure, *unless the independent brake-valve handle be placed in* **Lap Position** *before the automatic application is made.* If either section of this pipe is broken off, the effect will be the same as severe leakage.

Find the point of leakage from tests as suggested above. The temporary, road remedies are: If the *distributing-valve section* of this pipe is ruptured (be sure in cases of leakage from these small, copper pipes that it is not simply due to loose pipe-union nuts that can be quickly tightened, and the leakage stopped), plug the pipe toward the distributing valve, and disconnect a union in the double-heading pipe—preferably at a point in the cab; you will then have no use whatever of the independent brake-valve, but the locomotive brake will apply as usual from service or emergency reductions of brake-pipe pressure; the *holding effect* on the locomotive brake in **Release** and **Holding Positions** of the automatic brake-valve is lost, however, and this should be borne in mind when making all applications while running. If the automatic brake-valve section of this pipe is leaking or broken, keep going on: and, if you can't stop the leak, just before making an automatic application place the independent brake-valve handle in **Lap Position,** and the locomotive brake will operate, and hold, as usual; when the automatic brake-valve handle has been returned to **Running Position** (and not before), return the handle of the independent brake-valve to **Running Position,** and the locomotive brake will release.

If the Double-Heading Pipe (*blue*-colored, see Fig. 48) develops a leak, or breaks between the distributing valve and double-heading cock, the only effect is that it will be noticed that the locomotive brake will **release** in the **Release** and **Holding Positions** of the automatic brake-valve; all that is necessary to do in this case—if it is the leading engine, from which the train brakes are operated—is to plug the pipe toward the distributing valve, and no difference from normal brake operation will be experienced. With this section of pipe broken on one of the "*cut-out*" engines in double- or triple-heading, just pay no attention to it, for at this time that pipe has an atmospheric terminus anyhow; the only difference is that this engineer could not retain his locomotive brake in case of a train-brake release while running, made from the leading engine, except by placing his independent brake-valve in **Quick Application Position.**

If the section of this pipe between the double-heading

cock and automatic brake-valve is broken or leaking, it will have absolutely no effect on the engine from which the train braking is being done; but on all "*cut-out*" engines that may be in the train the result will be just the same as already explained in reference to the main section of this pipe connecting with the distributing valve.

Examination Questions and Answers

ON THE

E-T Locomotive-Brake Equipment

No. 5 and No. 6

Q. 1.—What differences are there in handle positions, and general operation, between the Engineer's Brake-Valves—Automatic, and Independent,—of the No. 5 and No. 6 styles, E-T locomotive-brake equipment?

A.—No difference ordinarily. On secondary engines in double heading, there is a difference in the positions in which the different automatic brake-valve handles should be carried.

Q. 2.—Name the Positions of the Handle of the Automatic Brake-Valve used in the E-T equipment.

A.—Beginning with the leftward, the positions are **Release, Running, Holding, Lap, Service-Application,** and **Emergency-Application.**

Q. 3.—What is the effect of the Release Position?

A.—In **Release Position,** main-reservoir pressure flows directly to the brake pipe, and, after an application, releases the train brakes, but the locomotive brakes remain applied. It does not affect the normal action of the pump governor. The warning port blows.

Q. 4.—What changes occur when the handle is moved to Running Position?

A.—The locomotive brake releases. The direct flow of main-reservoir pressure to the brake pipe is stopped; but its air, reduced by the feed valve to 70 pounds, is then supplied to the brake pipe, to the pressure chamber of the distributing valve, and the auxiliary reservoirs of the cars in the train. The pump governor remains unaffected.

Q. 5.—If the brake-valve handle had been brought from Release Position to Holding Position at once, what would have been the effect?

A.—The effect would have been the same as in **Running Position,** except that in **Holding Position** the locomotive brake would not release.

Q. 6.—After an automatic application of the brakes on the locomotive and cars of a short train, if the brake-valve handle is placed in Running Position, what will result? If placed in Holding Position?

A.—If the handle is placed in **Running Position,** the brakes on locomotive and cars will release; if in **Holding Position,** the car brakes only will release. In neither case will more than 70 pounds pressure flow into the brake pipe.

Q. 7.—What are the maximum main-reservoir and brake-pipe pressures possible in the first three brake-valve-handle positions just referred to, as commonly used?

A.—If the brake-valve handle had been left in **Release Position** long enough, main-reservoir and brake-pipe pressures would have equalized at 90 pounds; in **Running** and **Holding Positions,** main-reservoir pressure 90 pounds, and brake-pipe pressure 70 pounds.

Q. 8.—What results when the handle is placed in the Lap Position?

A.—All supply of air pressure to the brake pipe is cut off. The (90-pound) excess-pressure head of the pump governor is cut out of service, and the main-reservoir pressure will be increased to 110 pounds. This condition of pump-governor action, and pressure of main-reservoir air, is maintained, also, in both application positions.

Q. 9.—What is the effect when the brake-valve handle is placed in the Service-Application Position?

A.—Brake-pipe pressure is reduced: rapidly enough to cause the brakes of the locomotive and a train of any length to apply with service action, but not fast enough to cause quick action of any of the triple valves; the number of pounds-pressure reduction being indicated by the black hand on the large duplex gauge that registers "*equalizing-reservoir pressure*"; after the gauge has shown the desired amount of reduction, the handle should be returned to **Lap Position.**

Q. 10.—After such a graduated service reduction as alluded to, when the handle has been returned to Lap Position, will the discharge of brake-pipe pressure cease at once?

A.—Not necessarily. With a brake-pipe air volume no greater than that of the locomotive and one or two cars, the pressure discharge will cease as soon as the brake-valve handle is brought to the **Lap Position;** with more than that number of cars, the discharge of brake-pipe pressure will continue for a time after the brake valve has been lapped—the longer the train line, the longer will be the duration of the pressure discharge.

Q. 11.—What results when the brake-valve handle is placed quickly in the Emergency-Application Position?

A.—In the **Emergency-Application Position** the brake-pipe air is heavily discharged and its pressure quickly reduced, through a large port in the rotary valve, resulting in the almost instantaneous application of every cut-in brake in the train, the triple valves operating with *quick action.*

Q. 12.—What differences are there as between the Air Gauges of No. 5 and No. 6 E-T Locomotive-Brake Equipments?

A.—There are 2 air gauges in all styles of the E-T equipment; the larger is always a *duplex gauge,* which shows *main-reservoir pressure* by the Red Hand, and *Equalizing-Reservoir Pressure* by the Black Hand; in the No. 5 equipment, the dial of this gauge reads "Red Hand Main-Reservoir Pressure," and "Black Hand Train-Line Pressure"; while the No. 6 gauge reads the same as to the Red Hand, but states more correctly—"Black Hand Equalizing-Reservoir Pressure." The smaller gauge is of the single-pointer style in the No. 5 equipment, the hand is Black, and indicates "Locomotive Brake-Cylinder Pressure"; in the No. 6 equipment, it is of the duplex style, and the dial is changed to read—"Red Hand Brake-Cylinder Pressure," and "Black Hand Brake-Pipe Pressure."

Q. 13.—Name the Positions of the Handle of the Independent Brake-Valve.

A.—From the left, the Handle Positions are **Release, Running, Lap, Slow-Application,** and **Quick-Application.**

Q. 14.—Why would it be impossible to leave the handle of the independent brake-valve in Release, or Quick-Application, Positions?

A.—Because the *return spring* within the valve body will automatically rotate the rotary valve from **Release** to **Running Position,** and, in the No. 6 equipment, from **Quick-Application Position** to **Slow-Application Position.**

Q. 15.—What is the result when the independent brake-valve handle is put into the Release Position?

A.—The locomotive brake will release, after any manner of application whatever. In this position of the independent brake-valve of the No. 6 equipment, a *warning port* is caused to blow, as a measure of safety in case of a broken return-spring, as, if the independent brake-valve should remain in **Release Position,** it would be impossible to apply the locomotive brake in any manner.

Q. 16.—What is the effect of the Running Position of the independent Brake-Valve Handle?

A.—It is the regular carrying position for the brake-valve handle, and *must not be moved therefrom* except to apply the independent locomotive brake, or to release it when the automatic brake-valve handle is in some other than its Running Position. When the automatic brake-valve handle is in **Running Position,** and a locomotive-brake application has been made by the independent brake-valve, in order to release it it is only necessary to place the independent brake-valve in **Running Position.** The locomotive brake can not be released by the automatic brake-valve unless the independent brake-valve is in **Running Position.**

Q. 17.—What is effected in the Lap Position of this brake-valve?

A.—As in any other brake-valve, all ports in the rotary valve and rotary-valve seat that are separable, are closed; it is the negative position to which the handle is returned after making a graduated, independent application.

Q. 18.—Explain the Slow-Application Position.

A.—In this position the locomotive brake will be applied slowly, as the term indicates, giving the engineer the opportunity to graduate the application as finely as he desires. After a *graduated application*, the handle should be returned to the **Lap Position,** but when it is required that the locomotive shall be held for some time under the control of the independent brake, leave the handle in this position of **Slow Application.**

Q. 19.—Explain the Quick-Application Position.

A.—The action of all parts affected during an independent-brake application is no different as between the **Slow-Application** and **Quick-Application Positions,** except that in the latter position braking pressure is supplied to the engine- and tender-brake cylinders through a larger port in the rotary valve, giving, as the name implies, a *quick* action of the locomotive brake.

Q. 20.—What regulates the Brake-Pipe Pressure in the E-T equipment?

A.—The Feed Valve.

Q. 21.—What regulates the Main-Reservoir Pressure?

A.—The Duplex Pump-Governor.

Q. 22.—What pressure is supplied to the Independent Brake-Valve? What regulates it at that figure?

Also, what other air-pressure-using device is supplied from the same source?

A.—45 pounds pressure is supplied to the independent brake-valve, by the Reducing Valve, which also furnishes the pressure used in the Train Air-Signal System.

Q. 23.—Where does the pressure for the locomotive brake cylinders come directly from, at an automatic application? At an independent application?

A.—In both cases, from the main reservoir.

Q. 24.—In each case, what reduces the pressure, and regulates the amount?

A.—At an automatic application, the distributing valve, influenced by the amount of brake-pipe-pressure reduction; at an independent application, by the independent brake-valve, or the reducing valve.

Q. 25.—If we wish to carry 70 pounds brake-pipe pressure, and 90 pounds main-reservoir pressure, with brake-valves in Running Position, but after pumping up to the limit we have pressures of 60 pounds and 90 pounds, respectively; is the pump governor all right? What changes should be made to secure the desired pressures?

A.—Although 90 pounds is the pressure desired in the main reservoir, the governor is not adjusted correctly. The responsible governor top *is not expected to regulate the main-reservoir pressure at 90 pounds,* but to regulate that pressure at a figure 20 *pounds higher than that in the brake pipe.* As the case stands, the regulating spring of the excess-pressure governor top should be slackened until the gauge shows 80 pounds on the Red Hand, as against the 60 pounds on the Black Hand; then, slowly turn the hand-wheel on the feed valve clockwise, tightening the regulating spring, and both gauge hands will rise equidistantly until they stand as desired at 70 pounds brake-pipe, and 90 pounds main-reservoir, pressures.

Q. 26.—With these pressures secured, suppose that you should have to operate a High-Speed-Braked passenger train, what changes would you be required to make in the air-brake equipment?

A.—To change the E-T equipment from the common "70-pound brake" to the High-Speed Brake, is a very simple matter, indeed. When such change may be anticipated, the high-pressure governor top should be permanently adjusted at a figure some higher than 130 pounds—say 140 pounds; also, the high-pressure stop on the feed valve should be already adjusted and tightly clamped in the proper position. To make the change it is only necessary to revolve the wheel handle of the feed valve clockwise until the pin on the wheel strikes the stop situated diametrically opposite the 70-pound stop; brake-pipe and main-reservoir pressures will be equally and automatically advanced by this simple act, to 110 pounds and 130 pounds, respectively, and so maintained while the automatic brake-valve is in **Running Position**; but when the handle is moved to **Lap,** or either of the **Application Positions,** main-reservoir pressure will be increased to 140 pounds, as a measure toward prompt train-brake release.

Q. 27.—If the brake-cylinder Piston Travel becomes excessively long, on the locomotive or tender, will the force on the piston be reduced thereby, as it is in the ordinary automatic brake?

A.—No; the air pressure per square inch on the pistons will not be affected by variations of the piston travel, and the holding power will be the same for any given degree of application, so long as the piston does not strike the non-pressure (back) head of the brake cylinder; and the pressures per square inch will be equal in the cylinders of the driver, tender, and truck brake. With too long piston-travel, the brake will be tardy in completely releasing, however.

Q. 28.—What will be the effect of Leakage of Locomotive Brake-Cylinder Pressure?

A.—An amount of brake-cylinder-pressure leakage that could render the ordinary automatic brake absolutely ineffective will not at all weaken the holding power of the E-T brake, for in the latter this pressure is maintained —insured against leakage, or the loss of pressure from leakage.

Q. 29.—To preserve this feature of locomotive braking-pressure maintenance, is it not essential that certain other parts of the locomotive air-brake equipment shall be absolutely free from leakage to the atmosphere?

A.—Yes; the *two little copper pipes* that connect to the left side of the distributing valve must be perfectly free from leakage clear to their further terminals; also the head of the upper portion of the distributing valve (application-piston cylinder cover) must be tight.

Q. 30.—If, after an application by either brake-valve, when a release is attempted the locomotive brake-cylinder pressure will be exhausted until only a few pounds remains when its escape ceases, and the brake remains lightly "stuck"—this at a release by either brake valve—and the only way found to completely release it is to move the independent brake-valve handle to the full Application Position for a second, and then throw it quickly to Release Position,—what could be the cause of the trouble?

A.—The packing of Application Piston 10, in the upper portion of the distributing valve, is in bad condition and the Application and Exhaust Slide Valves lack lubrication, and are gummy. Experience has proven that the maintenance of proper condition of this piston is at once the most important feature, and the hardest thing to secure, of anything relative to the distributing valve.

Q. 31.—Where is the Double-Heading Cock, and in what way does it differ, constructively, as between the No. 5 and No. 6 equipments?

A.—The double-heading cock is placed in the automatic brake-valve branch of the brake pipe, and is located directly beneath the brake-valve. In the No. 5 equipment it is double ported, one port through it conveying brake-pipe pressure, and at an angle of 90 degrees to it is the port to connect with the double-heading pipe when the cock is in the "closed" position—turned 90 degrees, and closed to brake-pipe pressure. In the No. 6 equipment there is no "double-heading pipe," and the Double-Heading Cock is a common, 1-inch cut-out cock.

Q. 32.—In case of Double Heading, if the engineer of the leading engine operates the train and locomotive brakes, what shall he do specially under the circumstances if his engine has E-T equipment?

A.—He shall make no changes in the equipment in any way, and shall operate the brakes just the same as if there was no other engine in the train.

Q. 33.—Under the same circumstances what should the engineer of the following engine do, if he has the E-T equipment?

A.—He shall permit his air pump to run as usual;

shall close the double-heading cock. If he has the No. 5 equipment, he shall place his automatic brake-valve handle on **Lap,** as the regular carrying position for it under the circumstances. With the No. 6 equipment, the double-heading cock must be closed, but the automatic brake-valve handle should be left in the **Running Position.**

Q. 34.—Suppose the air pump on the second engine should be broken down; or a "dead" engine is being towed; in either case, with E-T equipment, would the locomotive brake on such engine be operative?

A. Not unless that locomotive was equipped with the parts supplementary to the regular E-T equipment called the "Dead-Engine Feature."

Q. 35.—Explain the Dead-Engine Feature, and the principle of its use.

A.—It comprises a branch pipe connecting the main brake pipe, or *train line,* with a conveniently located pipe containing main-reservoir pressure; this branch pipe contains a *cut-out cock* which is normally closed, a small cylinder filled with curled hair that acts as an *air strainer,* a *check-valve,* and a *choke fitting* with a small hole (about $\frac{1}{16}$-inch diameter) through it which limits the flow of air through the device (generally termed the "Combined Air Strainer and Check-Valve"). When a locomotive whose air pump is inoperative is coupled into a train the brakes of which are to be operated from another engine, the engineman on the "dead" locomotive (in addition to closing the double-heading cock and placing his automatic brake-valve handle as prescribed in answer to question 33) should open the cut-out cock in his dead-engine feature; this will charge his main reservoir to a pressure nearly equal to that of the brake pipe, and higher than is really necessary. The check-valve prevents back flow of main-reservoir pressure to the brake pipe when the operating engineer makes an automatic application, and the choke fitting limits the drain from the brake pipe to about the time flow through the feed groove of a triple valve in charging an auxiliary reservoir. The brakes of the "dead" engine will then be automatically operative, and its independent brake could be applied, if necessary.

Q. 36.—Is the same set of fixtures comprised in the dead-engine feature used in another connection, in the E-T locomotive equipment?

A.—Yes; if the Train Air Signal is used, the Westinghouse furnishings for the dead-engine feature are used to connect the Reducing-Valve Pipe (of 45 pounds pressure) with the main Signal Pipe.

Q. 37.—What differences, if any, are found in the Combined Air Strainer and Check-Valve as between its use in the Signal Line, and as the Dead-Engine Feature?

A.—In connection with the Signal Line, a light spring

is used over the check-valve, while in the Dead-Engine Feature the check-valve spring is much stiffer.

Q. 38.—At what Pressure should the Safety Valve be adjusted?

A.—The safety valve on the No. 5 equipment distributing-valve should be adjusted at 53 pounds; on the No. 6 distributing valve, it should be set at 68 pounds.

Q. 39.—When it is desired to inspect one of the brake-valves, to remove the feed valve, reducing valve, or distributing valve, or to make any other disconnection of parts in the E-T equipment, will it be necessary to shut down the air pump and blow off the main-reservoir pressure?

A.—No. This would only be necessary in case the high-pressure governor top, or the air pipe leading to it, had to be disconnected. Before making any disconnection in all other parts of the E-T equipment, place the automatic brake-valve handle in **Release Position** and close the Main-Reservoir Cut-out Cock; this is a sort of 3-way cock, and, in addition to cutting off the supply of main-reservoir pressure from the general equipment, the latter is drained of pressure through a small port in the cock. The direct air-passageway through the automatic brake-valve, made by placing the handle in **Release Position,** is to prevent the lifting of its rotary valve and of the slide valve in the feed valve, with the chances of dirt lodging on the seats of those valves when the air current is reversed in direction of flow.

Q. 40.—What is the upper portion of the Distributing Valve called? What is its duty?

A.—The upper portion of the valve section of the distributing valve is called the Application Portion; its contained mechanism is given the duty of supplying main-reservoir air to the locomotive brake-cylinders at automatic and independent applications; and of discharging the brake-cylinder pressure, at automatic or independent release.

Q. 41.—What is the lower portion of the distributing valve called? And what is its duty?

A.—The lower portion of the valve section of the distributing valve is called the Equalizing Portion; its contained mechanism acts only in response to reductions and recharges of brake-pipe pressure; resembling a triple valve in its operation, its duty is to actuate the upper, or Application, mechanism when an automatic application or release is initiated; to graduate the power of automatic applications.

Q. 42.—What is the reservoir section of the distributing valve called? And what duty does it perform?

A.—It is called the Double-Chamber Reservoir, being separated by an internal, air-tight dividing wall into two compartments—one large one, and a much smaller one; the larger compartment is termed the Pressure

Chamber, and the smaller one the Application Chamber. Together they represent the *auxiliary reservoir* and (dummy) *brake cylinder* as associate parts of the lower, or *triple valve*, portion of the valve section.

Q. 43.—What apparent difference is there between the Distributing Valves of the No. 5 and the No. 6 E-T equipments?

A.—In the head of the lower portion of the No. 6 distributing valve, there is an Equalizing-Piston Graduating Spring; while the No. 5 distributing valve contains no such Graduating Spring.

Q. 44.—What is the "Quick-Action Cylinder Cap," used in connection with the E-T equipment?—Explain its duty, and can it be used with either style of the E-T equipment?

A.—The Quick-Action Cylinder Cap is a special head for the lower, or equalizing, portion of the No. 6 distributing valve, *only;* and only furnished when specifically ordered. Its duty is to discharge a portion of the brake-pipe air when an automatic emergency-application is made; this air discharged from the brake pipe flashes into the locomotive brake cylinders before the upper, or application, mechanism has had time to open the brake-cylinder supply from the main reservoir.

Q. 45.—Does this effect the increase of locomotive brake-cylinder pressure at an emergency application, that is noticed on the gauge?

A.—No. The brake-cylinder pressure is governed by the distributing valve, no matter what its source of supply; it is greater at an emergency application than at a full service application, but the increase is due to *the reduced expansion of pressure-chamber air* at an emergency application.

Q. 46.—How should the brakes of a passenger train be operated in making regular Station Stops, with the E-T equipment on the locomotive?

A.—By the Two-Application method. This consists in making two separate service-applications: the first, quite heavy, and the second, light. After the speed of the train has been heavily reduced by the first application, the train- and locomotive-brakes should be released by placing the automatic brake-valve handle in **Running Position;** the second application should be released *just before the wheels stop turning*, by placing the brake-valve handle in **Release,** and then **Holding Position;** and if the track is not level, the handle may be left in the latter position until the signal to start is received: otherwise, return the handle to **Running Position,** as soon as the train is stopped.

Q. 47.—How should the brakes on a very long passenger train be operated?

A.—The brakes on a very long passenger train should be operated about the same as is recommended for freight-train braking, the amount of draft-gear slack and the long train-line, or brake piping, making the conditions and requirements nearly the same.

Q. 48.—**How should the brakes of a freight train be operated, with E-T equipment on the locomotive?**

A.—Where a stop is intended, hold the train- and locomotive-brakes on until the wheels stop turning. Make the initial reduction of brake-pipe pressure as heavy as the stopping distance, speed, and gradient will permit. If a slowdown, only, is made, release the train brakes by placing the automatic brake-valve handle in **Release Position,** afterward moving the handle to **Holding Position;** before returning it to its regular carrying position, give the handle a second push into **Release Position,** leaving it there but 2 or 3 seconds, and return to **Running Position.** Always use the **Release Position** to release the *train brakes.*

Q. 49.—**What particular instructions should be remembered and always be observed, concerning the Independent Brake-Valve?**

A.—When using the independent brake to hold the locomotive while standing on the turntable, at a water column, coal chute, etc., do not return the independent brake-valve handle to the **Lap Position** but *leave it in* **Slow-Application Position.** Be exceedingly cautious in applying the independent brake while running, when the locomotive is attached to a long freight train, as it can cause the slack to run in with a tremendous shocking effect. With a passenger train, do not apply and release the independent locomotive brake while the engine is using steam, just to "steady the train" while rounding curves at a good rate of speed. With any kind of train and under all circumstances while running at any rate of speed, use the independent brake-valve only when actually necessary, and then with the exercise of the best of judgment. Use it altogether in operating the locomotive brake when detached from a train.

Q. 50.—What style of Hose and Couplings are used in the Brake-Cylinder Pressure line between the engine and tender, in the E-T equipment?

A.—One pair of Signal Hose, complete (1-inch hose, with ¾-inch nipples, and signal couplings).

Q. 51.—If the main Brake Pipe (train line) should break off under the tender, how could the train brakes be operated?

A.—By using the signal pipe to convey the brake-pipe air past the tender. Cross-couple the brake hose from engine with signal hose of tender, and signal hose from the rear of tender with brake hose on head end of first car; open all cocks in the made-up line, and close the cut-out cock in the signal line on the engine. Make the cross-connections between the brake and signal hose-couplings with a "*combination hose*," or, if none such are carried, by forcing the different styles of hose-couplings together. In the latter case, after finishing the trip new hose should be applied, as forcing them together damages the couplings for regular service.

Q. 52.—After making this arrangement for getting around the burst brake-pipe under the tender, could the air signal then be used?

A.—No; it would then be inoperative from the cars.

Q. 53.—If the pilot branch of Brake Pipe is found broken when coupling to a train in front of engine, how then can it be arranged to operate the brakes of the train?

A.—First, couple the brake hose and signal hose together at rear of the tender; then couple the signal hose at pilot with the brake hose of the adjoining car; open all cocks in the made-up line, and close the supply cut-out cock in the signal line on engine.

Q. 54.—If the Equalizing-Reservoir Pipe breaks off, how should you remedy matters?

A.—Plug the broken pipe; also plug the service-exhaust fitting in the bottom of the automatic brake-valve; proceed, making service reductions by moving the automatic brake-valve handle, carefully, a short distance into the **Emergency-Application Position.**

Q. 55.—If the Main-Reservoir Supply Pipe to Distributing Valve becomes ruptured, what are the instructions in such cases?

A.—Plug the broken pipe toward the main-reservoir

pressure, or, if there is a cut-out cock in the piece of the broken pipe toward the main reservoir, close it; then proceed. Brakes on the train are then operative as usual, but not on the locomotive, by either an independent or automatic application.

Q. 56.—If the Brake-Pipe Branch to Distributing Valve is broken, what should be done?

A.—With the No. 5 equipment, plug the broken pipe toward the brake pipe, and also close the cut-out cock in the main-reservoir supply pipe to distributing valve, if there is one; in some of the older equipments of this style of brake there is no such cut-out cock, and in that case place a blind gasket in a union in the pipe, and proceed, but without the use of the locomotive brake. With the No. 6 equipment, plug the broken pipe, and go right along. In the latter case, while the locomotive brake will not respond to automatic brake-pipe reductions, it can be applied by the independent brake-valve, but which must be placed in the **Release Position** in order to release it.

Q. 57.—If the Brake-Cylinder Pipe gets broken off, what should you do?

A.—If the pipe is broken between the distributing valve and brake-cylinder cut-out cocks, close the cut-out cock in the main-reservoir supply pipe to distributing valve, or, in its absence, use a blind gasket in a union. Proceed, without the use of the locomotive brake. If a section of this pipe line gets broken between a brake cylinder and its individual cut-out cock, just close that brake-cylinder cut-out cock, and go on, with only the loss of that one portion of the locomotive brake.

Q. 58.—If the Feed-Valve Pipe Branch to Excess-Pressure Governor Top should get broken off, how would you overcome the effect on the pump governor?

A.—Plug the broken pipe toward the feed-valve pipe, and place a blind gasket in a union in the pipe from automatic brake-valve to excess-pressure governor top. Proceed, with pump under permanent control of the high-pressure governor top.

Q. 59.—Suppose that the Main-Reservoir-Pressure Pipe from the Automatic Brake-Valve to the Excess-Pressure Governor Top breaks, en route, how would you get around this trouble?

A.—Plug the broken pipe toward the brake-valve, and proceed, with air pump under permanent control of the high-pressure governor top.

Q. 60.—Suppose that the direct Main-Reservoir-Air Pipe to High-Pressure Governor Top should be the broken one: What are the instructions in such case?

A.—Plug the broken pipe toward the main reservoir, and go on. Whenever an application is made by the automatic brake-valve, watch the main-reservoir-pressure gauge hand, and if it begins to rise too high, throttle the steam supply to the air pump.

Q. 61.—If the Reducing-Valve Pipe breaks, what should be done in the way of preventing it being the cause of other trouble, as well as stopping the escape of reducing-valve pressure?

A.—First, to stop the waste of air, slacken the pressure-adjusting nut of the reducing valve until no more air will feed from it; then, plug the broken pipe toward the independent brake-valve, and plug the exhaust port in the bottom of the independent brake-valve. Proceed, with the automatic brake operative as usual on the locomotive and train, but without the train air signal and *independent* locomotive brake.

Q. 62.—In the No. 5 equipment, suppose that the Double-Heading Pipe should get broken between the double-heading cock and the distributing valve: How would you remedy this case?

A.—The remedy is simply to plug the broken pipe toward the distributing valve; then go on. Everything will operate as usual.

Q. 63.—If, in the No. 5 equipment, the Application-Chamber Pipe should get broken, what should be done?

A.—Plug the pipe toward the distributing valve, and disconnect the double-heading pipe somewhere between the double-heading cock and distributing valve. The locomotive- and train-brakes can then be applied by the automatic brake-valve as usual, but the ordinary retaining feature of the locomotive brake in

Release and **Holding Positions** of the automatic brake-valve will be lost. The independent brake-valve will be inoperative, and should be left unused in the **Running Position.**

Q. 64.—In the No. 6 equipment, how would you overcome the effect of a broken Application-Cylinder Pipe?

A.—Plug the broken pipe toward the distributing valve, and proceed. The locomotive- and train-brakes can then be operated as usual by the automatic brake-valve, but the independent brake-valve will be out of service, completely.

Q. 65.—In the No. 6 equipment, suppose the Distributing-Valve Release Pipe should get broken off: what are the instructions?

A.—Keep going; but without the retaining, or holding, effect on the locomotive brake when the automatic brake-valve is placed in **Release** or **Holding Positions.** The locomotive brake can be applied by the independent brake-valve, but will release if that brake-valve is returned to the **Lap Position.**

INDEX

Index

C

Index

D

Index

E

F

G

H

I

L

M

N

O

P

Q

R

THE
WALSCHAERT
LOCOMOTIVE
VALVE GEAR
WOOD

Spotted something that isn't quite right?

Let us know and we'll fix it for future reprints. And as a thank you for your diligence, we'll send you a free Sothis book of your choice.

publisher@sothis.press

This page intentionally left blank

Made in the USA
Monee, IL
08 November 2026

68113593R00166